柏濤

建筑设计作品

2010—2023

PORTFOLIO OF PT ARCHITECTURE DESIGN

PT ARCHITECTURE DESIGN (SHENZHEN) CO., LTD.
柏涛建筑设计（深圳）有限公司 编

華中科技大學出版社
http://press.hust.edu.cn
中国 · 武汉

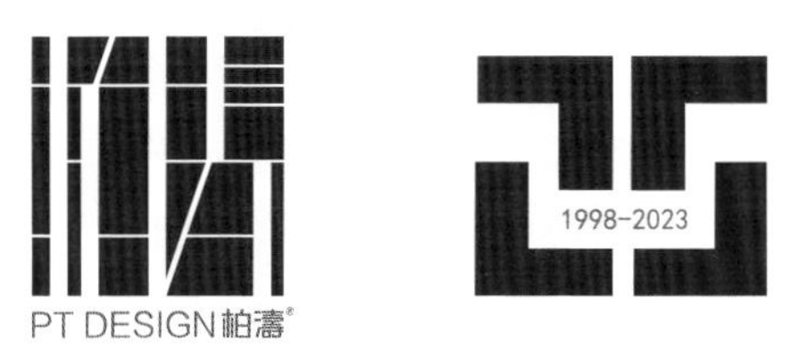
PT DESIGN 柏濤
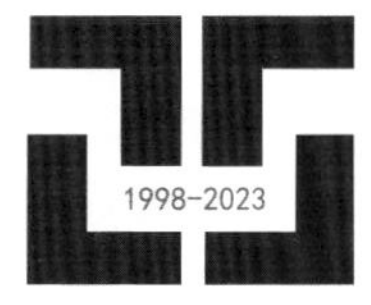
1998-2023

柏涛建筑设计 1998 年进入中国落地深圳，恰逢中国进入房地产真正市场化的新时代，25 年来见证了房地产行业从起步到高速发展到现阶段的高品质发展，也亲历了过程中起起伏伏的骤变。

2000 年至今是我国城镇化进程高速发展的 20 年。随着城镇人口的逐年增长，居住社区成为城市构成的重要组成系统之一，其发展也在一定程度上影响着一个城市的整体结构。“居者有其屋”是中国老百姓最朴实的愿望，居住社区和住宅承载着普通人的生活梦想，柏涛建筑设计以“设计创造美好生活”为己任，为中国的居住社区规划和住宅设计带来了很多的创新。《柏涛建筑设计 2010—2023 作品集》展现出柏涛建筑设计坚持在高品质居住区规划与住宅设计领域深耕创作的同时，拓展了其它建筑类型，比如体育公园、文化旅游、产业商业、教育办公等等，由此看到了当下高密度城市韧性、多元、复合发展的轨迹，看到了居民生活从物质空间到精神空间的追求，“美好生活”的范畴越来越广、越来越深 , 也看到了柏涛建筑设计在专业领域的专注和成长。

柏涛建筑设计在成立 25 周年之际出版这一册作品集既是沉淀，也是总结。期待下一个 5 年、10 年看到更优秀的柏涛设计，为建筑行业奉献更多的建筑佳作。

PT Architecture Design entered China in 1998 and landed in Shenzhen, which coincided with the cancellation of the real estate allocation policy in China and opened a new era of real estate marketization. In the past 25 years, we have witnessed the real estate industry from its inception to its rapid development and then to its present stage of high-quality development, and have also experienced the ups and downs of sudden changes in the process.

Since 2000, it has been 20 years of rapid urbanization in China. With the annual growth of urban population, the residential community has become one of the key compositions of the overall urban structure. Its development also affects the overall structure of a city to some extent. "Anyone who needs to settle down (life) should get a residence (house)" is the simplest wish of Chinese people, so living communities and residences bear the life dreams of ordinary people. PT Architecture Design has brought a lot of innovation and creativity to China's residential community planning and residential design with the "design to create a better life" as the concept and responsibility. *In the Portfolio of PT Architecture Design 2010-2023*, PT insists on deep creation in the field of high-quality residential planning and residential design and also expands other building types, such as sports parks, cultural tourism, industrial office, and school buildings. Through this, we can see the toughness, diversity, and compound development track of high-density cities, the residents' pursuit of life from material space to spiritual space, and the scope of "a good life" is getting wider and deeper. We also see the concentration and growth of PT Design in the professional field.

PT Architecture Design publishes this collection of works in the establishment of the 25th anniversary as both a precipitation and a summary. We look forward to seeing a more excellent PT Design in the next 5 years and even 10 years and hope PT Design will dedicate more excellent architectural works to the architecture industry.

孟建民

25 年，1/4 个世纪，一部沉甸甸的作品集，它浓缩了我们的奋斗精神、创造力和闪光的智慧，更浓缩了我们的岁月和青春！

当我们穿过时光隧道从 20 世纪末重新出发，见证了中国城市化的发展历程，也见证了中国城市沧桑巨变的点点滴滴，作为一线参与者，中国城市化的轨迹上有我们的签名，我们感到自豪和骄傲。

25 年前带着澳大利亚百年的设计品牌“Peddle Thorp”来到中国，和一帮因设计理想聚集在一起的伙伴们，以超前的设计理念和设计手法，开创了中国居住社区发展的新时代；同时也激发了具有相同使命感的同行为中国城市化变革发展进程贡献他们的智慧和才华！25 年间，中国的城市空间发生了翻天覆地的变化，25 年来柏涛建筑设计也经历了不同的发展时期，并取得了令人瞩目的成就。作为创始人，我感到自豪！

柏涛建筑设计的成功和发展离不开众多业内同行和广大客户的信任和支持，我们以“设计创造美好生活”为核心设计理念，一直坚持创新和追求卓越，不断提升服务品质。深得客户的信赖和好评。在此，我向所有支持我们的朋友表示诚挚的谢意和敬意！

“柏涛建筑设计，再行百年”不仅是我们的愿景和目标，也是我们的使命和责任。在这个充满变革和不确定性加剧的时代，我们将始终保持对设计的热情和对创意的追求，以高品质的设计和探索精神，秉持着对未来的信心和期许引领建筑设计行业的发展，不断创造令人瞩目的作品。感谢所有为柏涛建筑设计的发展做出贡献和付出的同仁们！

让我们通过柏涛建筑设计：见证历史！见证时代！

The works selected in this profound collection span 25 years, concentrating our struggle, creativity, and sparkling wisdom in the past quarter-century, and even our years and youth.

This collection will take us back to the end of the 20th century, to witness the development of China's urbanization and the vicissitudes of the changes in Chinese cities. As a front-line participant, we are proud and honored to have our signature on the trajectory of "China's urbanization", the largest urbanization development in human history.

25 years ago, I returned to China from Australia with Peddle Thorp, a century-old Australian design brand, and with a group of partners gathered for design ideals, we created a new era for the development of residential communities in China with cutting-edge design concepts and techniques; and this also inspired my peers with the same sense of mission to contribute their wisdom and talent to China's urbanization reform and development. In these 25 years, China's urban space has undergone tremendous changes, and PTAD has also experienced different periods of development and made remarkable achievements. As the founder of PTD, I am proud of us.

PTAD's success and development are inseparable from the trust and support of many industry peers and clients. We adhere to the core design concept of "design for a better life", insist on innovation and the pursuit of excellence, and constantly improve the quality of our services. Through unremitting efforts, we have been trusted and praised by our clients. Here, I would like to express my sincere gratitude and respect to all those who have supported and helped us.

For PTAD, another 100 years of development is not only our vision and goal but also our mission and responsibility. In this era of change and heightened uncertainty, we will always maintain our passion for design and the pursuit of creativity, leading the development of the architectural design industry with high-quality design and the spirit of exploration, upholding our confidence and expectations for the future, to continuously create remarkable works. Thanks to all peers and colleagues who have contributed to the development of PTAD.

Let us, through PTAD, witness history and witness the times!

柏涛建筑设计创始人：王漓峰
Wang Lifeng
Founder of PT Architecture Design

CONTENTS
目录

深圳 · 招商双玺时光道
IMPERIAL PARK OF CMPD · SHENZHEN

位　　置：广东省深圳市
客　　户：深圳招商房地产有限公司
用地面积：4.28 万 m^2
建筑面积：12 万 m^2
功　　能：高端住宅

Location: Shenzhen, Guangdong Province
Client: Shenzhen Merchants Property Development Co., Ltd.
Site Area: 42 800 m^2
Building Area: 120 000 m^2
Function: High-end residence

山之阶，海之纹
THE STAIRS OF THE MOUNTAIN, THE RIPPLES OF THE SEA

以传承海运文化为出发点，通过内敛克制的设计手法，意图最大程度回馈深圳蛇口海上世界片区。

Taking the inheritance of shipping culture as the starting point, this Project is intended to give back to Shekou Sea World Area in Shenzhen to the greatest extent through restrained design techniques.

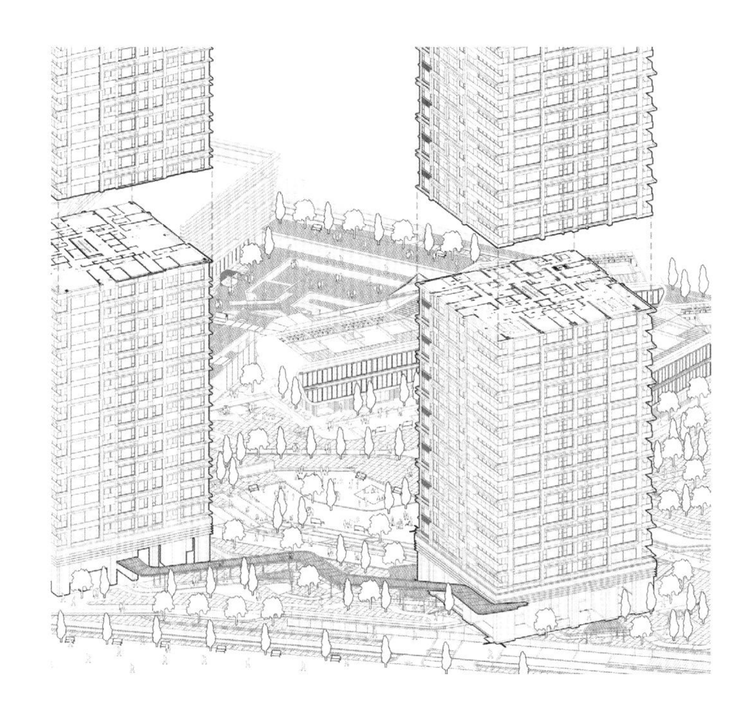

项目位于深圳市南山区蛇口，一个完整的、全过程的绿色化设计、科技化建造、人文化运营的滨海居住项目。

规划形态集约为 4 栋塔楼、4 栋洋房及 1 栋幼儿园，空间流动，通透性强，城市关系友好，通透性符合当地气候特征；通过遮阳、通风、室内光环境技术设计，使整个项目均能达到优良的微气候效果。

This Project is located in Shekou, Nanshan District, Shenzhen, which is a complete and whole-process coastal residential project with green design, technological construction, and humanistic operation.

This Project adopts an intensive planning form, which consists of four towers, four bungalows, and a kindergarten. With moderate and permeable spatial flow, and friendly urban relations, this Project is constructed in line with the local climate characteristics; through the technical analysis of shading, ventilation, and indoor light environment, the Project as a whole has been proven to be able to achieve excellent micro-climate effect.

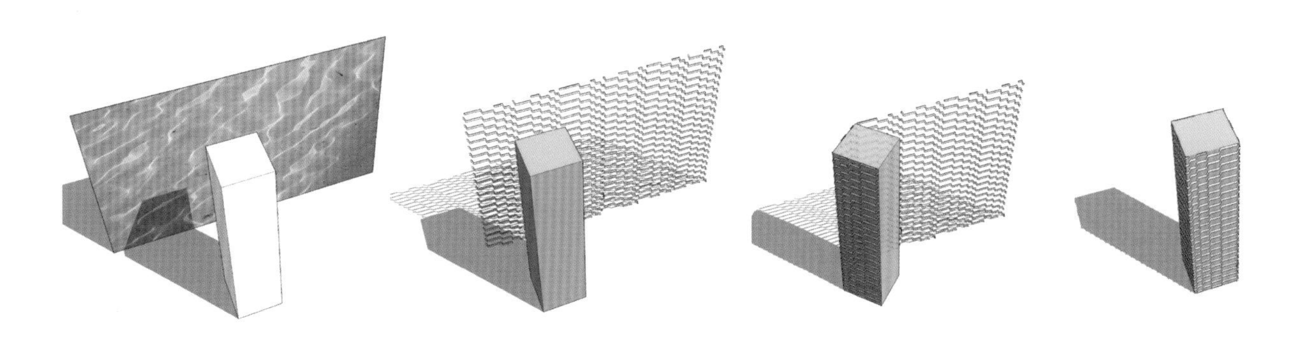

我们认为城市滨海特征是贯穿于该项目空间组织、平面、立面等各个设计层面的原则，应是一个兼具建筑设计的严谨与滨海生活放松浪漫于一体的存在。

立面设计构成元素提取了海浪、卵石等滨海符号，营造出水落石出的浪漫效果。

We always believe that the urban coastal character should be one of the principles running through the spatial organization, graphic design, elevation design, and other design levels of the Project, and this Project is designed to combine the rigor of architectural design with the relaxation and romance of coastal life.

As for the composition elements of the elevation design, we extracted coastal symbols such as waves and pebbles to create a romantic effect that as the water recedes, the stones appear.

南京·正荣滨江紫阙
ZHENRO BINJIANG ZIQUE · NANJING

位　　置：江苏省南京市
客　　户：南京正荣房地产开发有限公司
用地面积：4.43 万㎡
建筑面积：16.08 万㎡
功　　能：住宅

Location: Nanjing, Jiangsu Province
Client: Zhenro Properties Group (Nanjing) Co., Ltd.
Site Area: 44 300 m^2
Building Area: 160 800 m^2
Function: Residence

健康筑就品质的绿色住宅
HEALTHY AND HIGH-QUALITY GREEN RESIDENCE

项目围绕着“健康、绿色”的目标，在节能、环保等方面进行了一系列尝试，创造出了一个绿色、低能耗、宜居的社区环境。

Centering on the goal of "health and green", this Project has made a series of attempts in energy saving and environmental protection, creating a green, low energy consumption and livable community environment.

建筑规划错落有致，形成了“区中有园，园中有院，院中有落”的环境格局，既体现了中国传统的礼仪秩序，又利于环境中自然风的疏导。

At the planning level, the buildings are well-arranged, forming an environmental pattern of "gardens in the residential community, courtyards in the gardens, and small courtyards in the courtyards", which not only reflects the traditional Chinese etiquette order, but also facilitates the natural wind diversion of the environment.

立面形式采用现代主义风格，重视功能和材料的工艺特性，推崇科学合理的构造工艺。装配式技术的应用、工厂化生产为主的施工方式以及大量低碳技术的应用为本项目营造了舒适健康的居住环境，取得了良好的社会影响力和绿色示范效应。

The facade form adopts a modernist style, attaches importance to the technological characteristics of functions and materials, and respects scientific and reasonable construction techniques; the application of assembly technology, factory production-based construction method, and the application of various low-carbon technologies have created a comfortable and healthy living environment for this Project, which has also achieved considerable social impact and green demonstration effect.

深圳·深国际颐城栖湾里

YICHENG QIWANLI OF SHENZHEN INTERNATIONAL · SHENZHEN

位　　置：广东省深圳市
客　　户：深国际前海投资管理（深圳）有限公司
用地面积：2.53 万㎡
建筑面积：10.47 万㎡
功　　能：住宅、商业、幼儿园、公共配套

Location: Shenzhen, Guangdong Province
Client: Shenzhen International Qianhai Investment and Management (Shenzhen) Co., Ltd.
Land Area: 25 300 m^2
Building Area: 104 700 m^2
Function: Residence, Commerce, Kindergarten, Public Supporting

与自然相融的花园社区

A GARDEN COMMUNITY THAT INTEGRATES WITH NATURE

设计从城市空间界面整体形象、居住体验等多个维度进行思考，以“行云流水”作为设计灵感，串山连海，整合区域脉络，创造了一个多维渗透的绿色生态社区。

The design of this project is considered from many dimensions, such as urban space, interface image, and living experience, and takes "floating clouds and flowing water" as the inspiration, connecting mountains and seas, integrating regional context to create a multi-dimensional green ecological community.

项目底层空间设计结合绿色、开放、共享的概念，引入骑楼、挑檐等城市灰空间，创造出一个 24 小时立体公园生态街区。在商业重要节点引入“树状”造型，为水廊道沿线带来一道独特的风景线。塔楼立面设计通过“花瓣”造型，结合形体转角微曲元素，打造出一片自然和人文互相平衡的生态栖息地。

The design of the ground floor space of this Project combines the concepts of green, openness and sharing, and introduces urban gray spaces such as an arcade and overhanging eaves to create a 24-hour three-dimensional park ecological block. At key commercial nodes, this Project introduces a growing tree shape to create a unique landscape along the water corridor, while the facade design of the towers creates an ecological habitat where nature and humanity are in balance through the vertical "petal" shape and the combination of corner micro-curved elements.

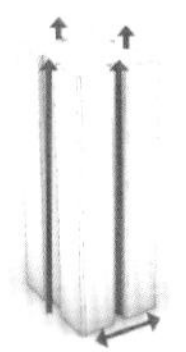

打破体量，视觉上竖向延伸

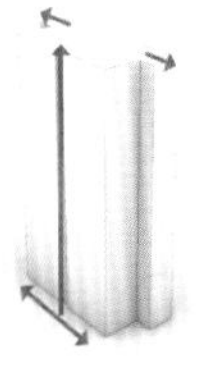

打破体量，视野最大化

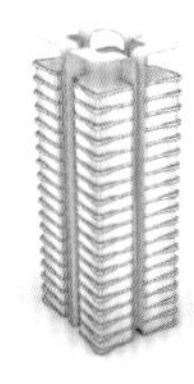

激活屋顶

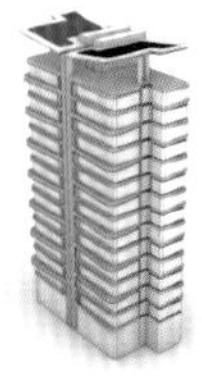

激活屋顶

深圳·滨江爱义南方大厦

BINJIANG AIYI NANFANG BUILDING · SHENZHEN

位　　置：广东省深圳市
客　　户：深圳市南方远大传媒置业有限公司
用地面积：0.56 万 m^2
建筑面积：6.7 万 m^2
功　　能：超高层办公、还迁住宅、商业

Location: Shenzhen, Guangdong Province
Client: Shenzhen Nanfangyuanda Media Real Estate Co., Ltd.
Land Area: 5600 m^2
Building Area: 67 000 m^2
Function: Super High-Rise Office, Relocated Residence, Commerce

城市叠合，打造生机勃勃的“城市之心”

CREATE A VIBRANT "URBAN CORE" THROUGH URBAN SUPERPOSITION

设计将办公、居住、商业、文化活动、社区交往、绿地系统叠合在一个立体的垂直空间中，通过“城市叠合”理念，打造区域生机勃勃的“城市之心”。

Through the concept of "urban superposition", the design of this Project integrates office, residential, commercial, cultural, community and green space functions in a three-dimensional vertical space, building a vibrant "urban core" of the region.

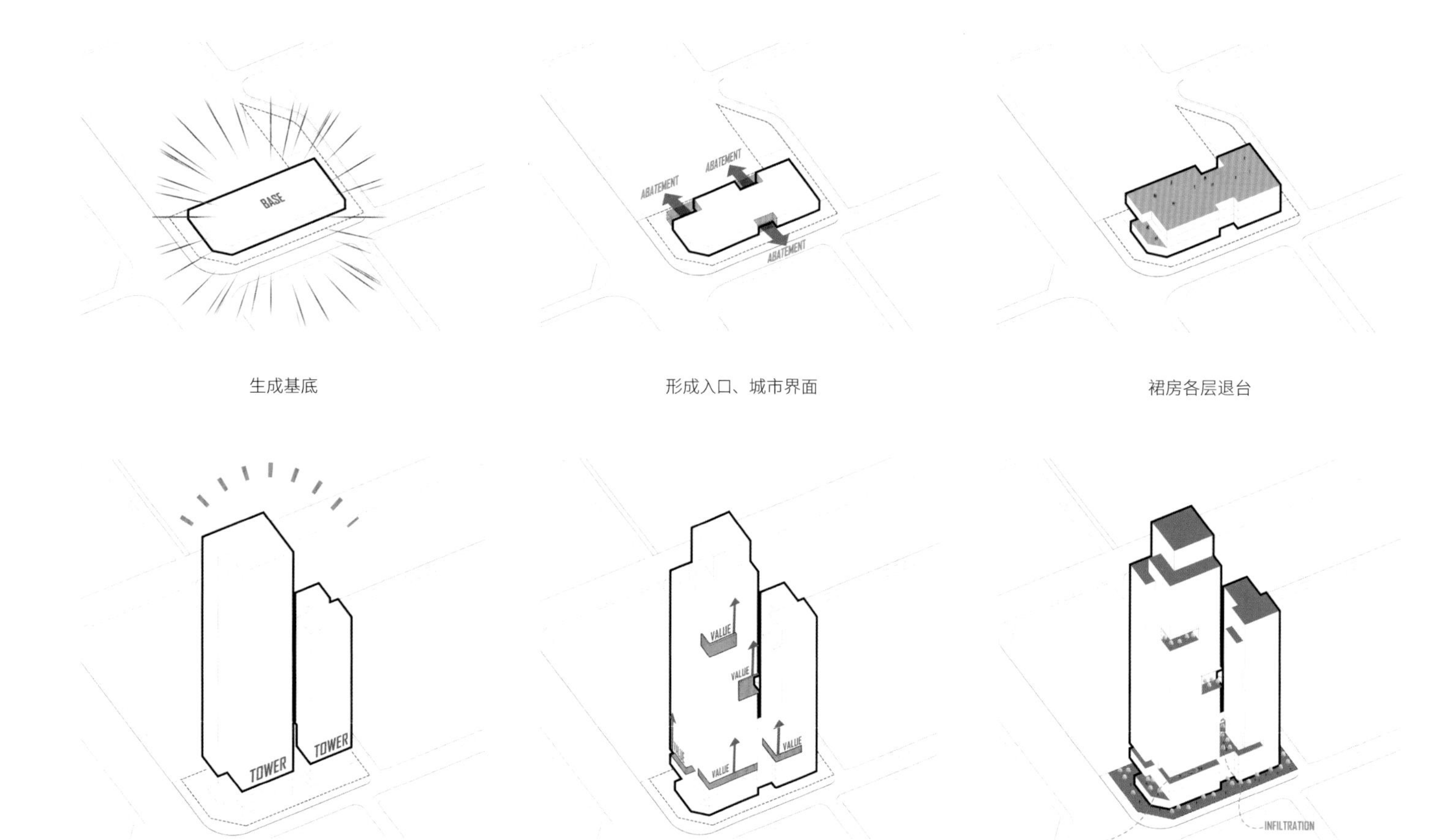

生成基底　　形成入口、城市界面　　裙房各层退台

放置塔楼　　价值提升点　　多层次绿化

项目力求在深圳核心区打造一个高度复合且与城市有机融合的片区标杆。设计在形体刻画上突破传统框架，通过不同高度的体量错动，形成从下沉广场到空中花园、从公共平台到屋顶退台的一系列多元立体的开放空间及绿色交往平台。

设计将建筑与 5G 时代的智慧交互及文化的传承，转译为干净利落的建筑语言，通过现代创新的设计手法，创造出开放的互动场所，打造舒适的人文环境及宜居的城市空间，最终浓缩成区域中生机勃勃的“城市之心”。

This Project aims to create a highly complex and organically integrated urban benchmark in the core area of Shenzhen. The design breaks away from the traditional framework in terms of form and structure and formulates a series of multi-dimensional open spaces and green interaction platforms, from sunken squares to sky gardens, from podium deck to roof terrace, through the staggering of building volumes at different heights.

The design translates the interaction between the architecture and the wisdom of the 5G era and the heritage of culture into a smooth and clean architectural language, creating an open and interactive place, a comfortable cultural environment, and a liveable urban space through modern and innovative design techniques, ultimately condensing it into a vibrant "Heart of the City" of the region.

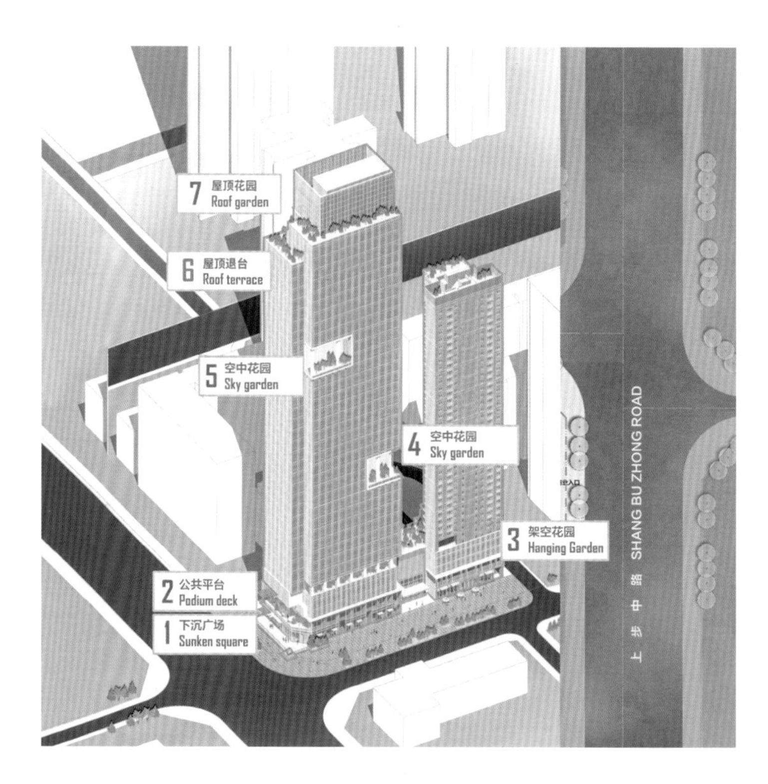

深圳 · 帝豪金融大厦
REGENCY FINANCIAL TOWER · SHENZHEN

位　　置：广东省深圳市
客　　户：深圳市顺豪实业有限公司
用地面积：0.5 万 m^2
建筑面积：8.47 万 m^2
功　　能：办公、商业、酒店、公寓

Location: Shenzhen, Guangdong Province
Client: Shenzhen Shunhao Shiye Co., Ltd.
Site Area: 5000 m^2
Building Area: 84 700 m^2
Function: Office, Commerce, Hotel, Apartment

云上舞者
DANCER ON THE CLOUDS

约 200 m 的云上舞者，以开放、复合、集约、浪漫为设计理念，为高密度背景下微小地块城市更新设计做了一次有益的探索。

The total height of the Project is about 200 meters, and its appearance is elegant, like a dancer on the clouds. The Project adopts an open, complex, intensive, and romantic design concept, which makes a beneficial exploration for the urban renewal design of small plots in high-density urban spaces.

万通大厦
WANTONG BUILDING

项目为一栋超高层综合楼，功能包含办公、商业、酒店、公寓等多种不同业态，创新的“Z 型”核心筒设计，让平面形成了大面宽小进深的平面格局。该模式打破了传统的大进深小面宽对于高品质居住的束缚，为小面积段功能的拼合创造了有利的条件，使得整栋建筑楼都拥有最大的外部景观及视线延展面，较好的解决了多种复合业态的不同功能性需求。

塔楼分成南北两翼成交错之势，“化整为零”的设计手法加强了建筑的竖向挺拔感，竖向线条加上局部的浪漫收分，形成了独特而又富有标志性的建筑形象，底部开放的公共空间似乎将建筑悬浮而起，宛若多姿的舞者，在云中绽放。

The entire project is a super high-rise complex building, with functions including office, commerce, hotel, and apartment, etc. The innovative "Z-shaped" plan forms a large width and small depth plan pattern, which not only breaks away from the traditional fetters of large depth and small width for high-quality living, but also creates favourable conditions for the combination of functions of small area sections, and at the same time makes the whole building have the largest external landscape and sight line extension, which better solves the different functional needs of multiple complex businesses.

The tower is divided into north and south wings in an interlaced manner, and the design technique of "dividing the whole into parts" greatly enhances the vertical uprightness of the building. The vertical lines, together with the local romantic closure, form a unique and iconic image of the building, and the public space at the bottom, which is open to the city, levitates the building from the ground, like a dancer on the clouds.

杭州·李宁体育园
LI-NING SPORTS PARK · HANGZHOU

位　　置：浙江省杭州市
客　　户：杭州市江干区艮山东路北区块开发建设指挥部办公室 / 杭州海胜建设有限公司
用地面积：5.76 万 m^2
建筑面积：6.88 万 m^2
功　　能：体育建筑

Location: Hangzhou, Zhejiang Province
Client: Development and Construction Headquarters Office of the North Block, Genshan East Road,
Jianggan District, Hangzhou / Hangzhou Haisheng Construction Co., Ltd.
Site Area: 57 600 m^2
Building Area: 68 800 m^2
Function: Sports Building

踏着云朵去奔跑
RUN WITH THE CLOUDS

设计采取大体量建筑地景化 + 高架跑步道策略，将景观和建筑串联起来，实现低覆盖率下的体育综合体设计，构建起城市空间、体育文化及居民健康生活的连接。

The design of this Project adopts the strategy of large volume building landscaping and elevated running track to link the landscape and the building to realize the design of the sports complex under low coverage and build a connection between the city and sports and then people's health.

在保证建筑满足多元使用功能的前提下，最大程度地利用、串联用地内公园绿地及室外运动场地资源，为市民提供一个高品质、高舒适度的体育锻炼和休闲娱乐场所，促进体育运动的普及和全民健身事业的发展。

设计结合建筑功能要求，根据杭州气候条件，遵循可持续发展原则，采用被动式能源策略，最大限度利用再生能源。场馆空间从能量储存、循环再利用出发，严选建筑材料。部分建筑空间的设计具有灵活可变性，以便减少建筑体量，将建设所需的资源降至最少。

Under the premise of ensuring that the building meets multiple functions, this Project aims to maximize the use and linkage of parkland and outdoor sports resources within the site to provide a high-quality and comfortable place for sports and recreation for the public, thus promoting the popularity of sports and the development of national fitness.

The design incorporates the functional requirements of the building and follows the principles of sustainable development according to the local climatic conditions of Hangzhou, and adopts a passive energy strategy to maximize the use of renewable energy. The venue spaces are designed from energy storage and recycling as a starting point, and building materials are strictly selected. Some of the building spaces are designed with flexibility and variability to reduce the building volume and minimize the resources required for construction.

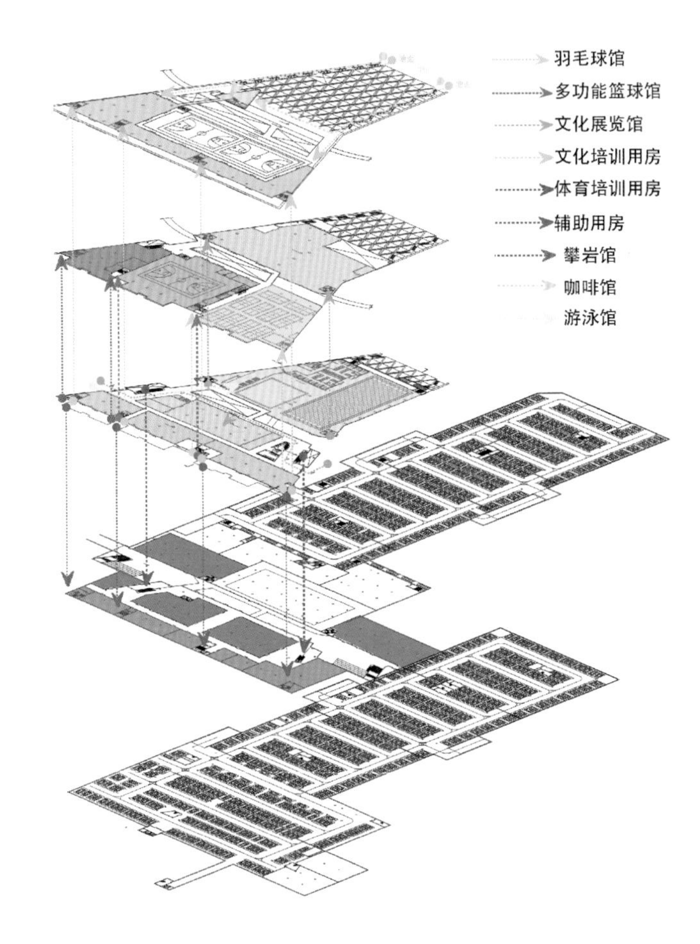

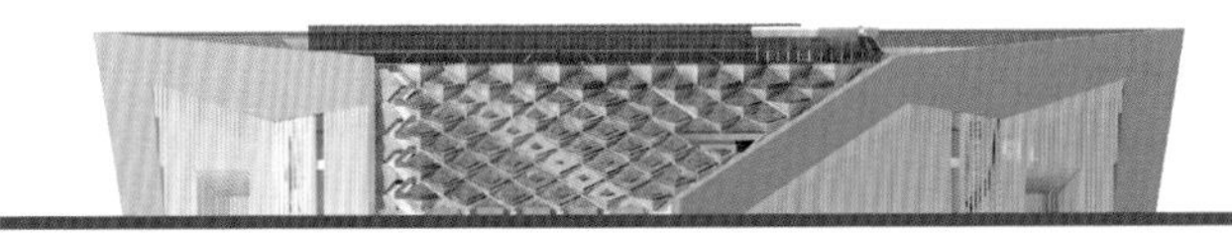
南立面图

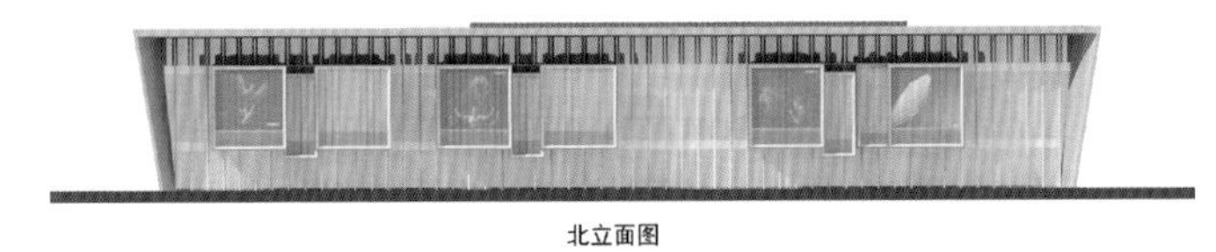
北立面图

东立面图

西立面图

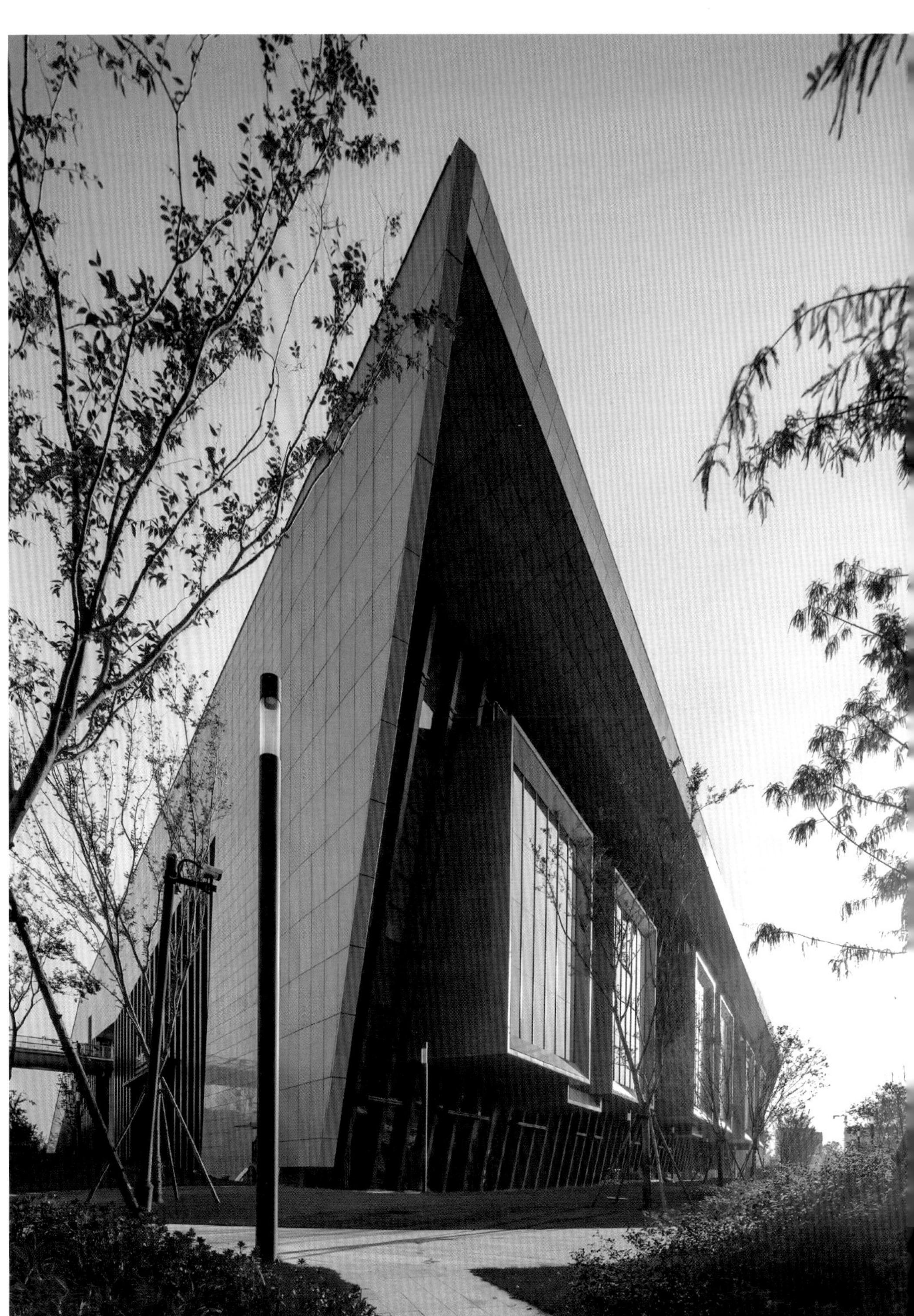

桐乡·李宁体育园
LI-NING SPORTS PARK · TONGXIANG

位　　置：浙江省桐乡市
客　　户：桐乡市振东新区建设投资有限公司
用地面积：15.27 万 m^2
建筑面积：9.64 万 m^2
功　　能：体育建筑

Location: Tongxiang, Zhejiang Province
Client: Tongxiang Zhendong New District Construction Investment Co., Ltd.
Site Area: 152 700 m^2
Building Area: 96 400 m^2
Function: Sports Building

体育运动催化城市生活
ACTIVATE A DIVERSE URBAN LIFE THROUGH SPORTS

设计引入"体育 MALL"的概念，采用了"一心多核"的放射式布局，将其整合成一个整体。

The design of this Project introduces the concept of "Sports Mall" and adopts a radial layout of "one center with multiple cores" to integrate the Sports Park into a whole.

设计的出发点是将建筑体量和体育公园充分融合成为一个整体，以健身中心为锚点，创造更多公共开放交流的空间，成为市民喜闻乐见的城市客厅。

空间规划方面，引入“体育 MALL”的概念，采用了“一心多核”的放射式布局。

设计以一条富有动感的连续曲面，围绕着室内室外一系列的活动内容组织空间关系，从城市到公园再到健身中心，形成独特的休闲体育文化体验流线。

我们希望，桐乡全民健身中心的建设不仅为桐乡及周边市民提供一个富有活力的公共活动场所，也为中国建设未来的群众体育设施提供一个有益的典范。

The starting point of the design is to fully integrate the building volume and the Sports Park in which it is located into a whole, taking the fitness center as an anchor point to create more public open communication spaces, seeking to become an enjoyable urban living room for citizens.

In terms of spatial planning, we introduce the concept of "Sports Mall" and adopts a radial layout of "one center with multiple cores".

The design features a dynamic and continuous curved surface that organizes spatial relationships around a series of indoor and outdoor activities, from the city to the park and then to the fitness center, forming a unique flow of sports and cultural leisure experiences.

We hope that the Tongxiang Li-Ning Sports Park is expected to provide a vibrant public activity venue for the citizens of Tongxiang and its surrounding areas, and also to provide a useful model for building future mass sports facilities in China.

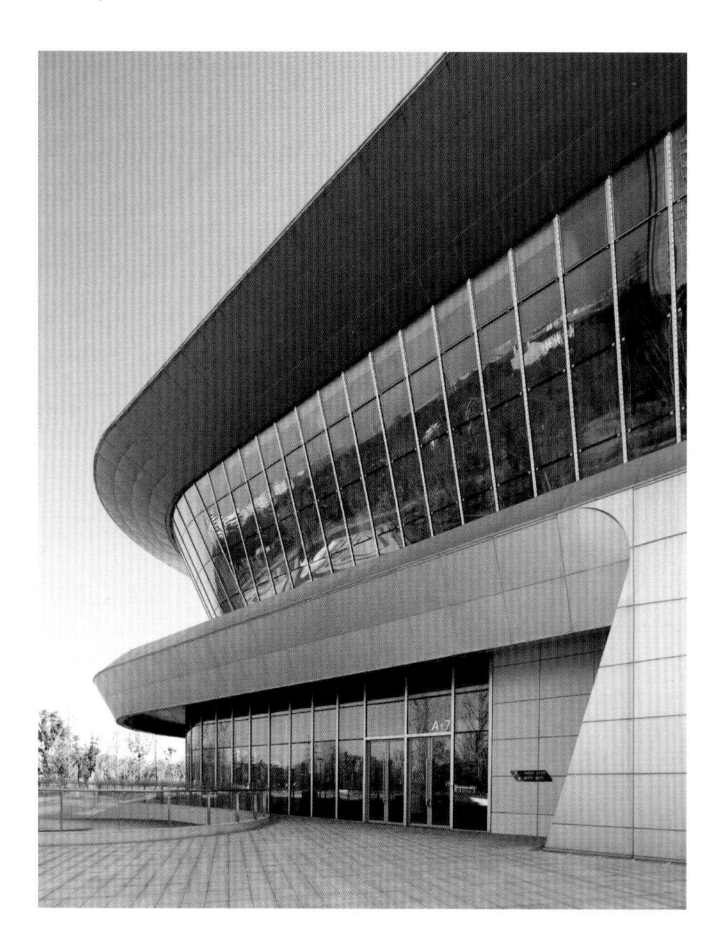

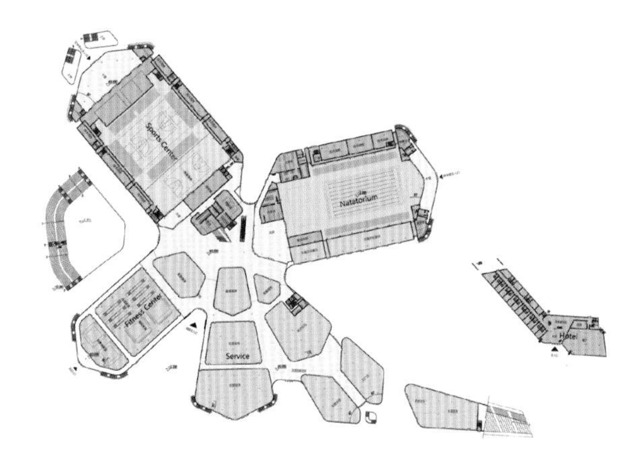

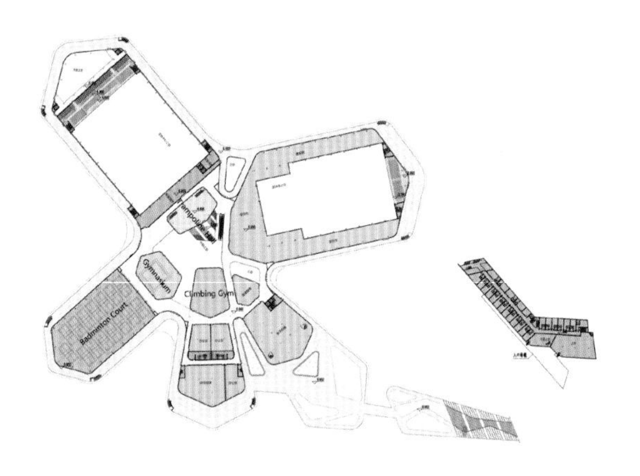

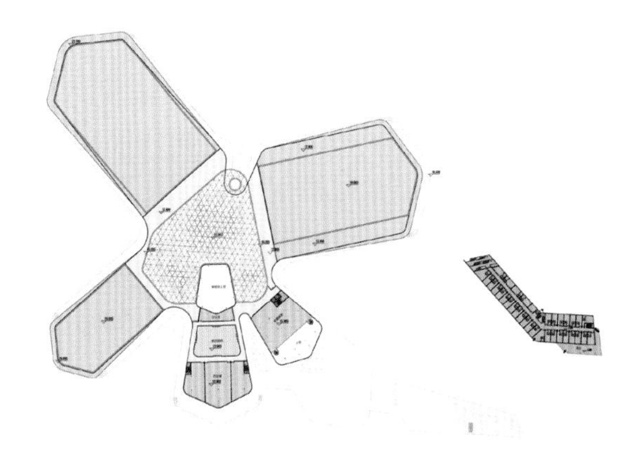

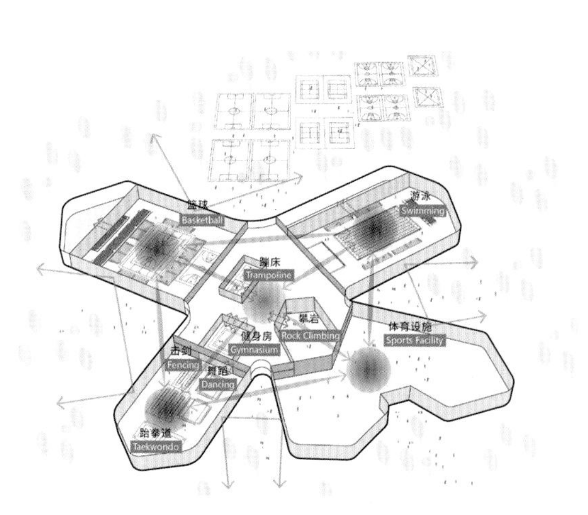

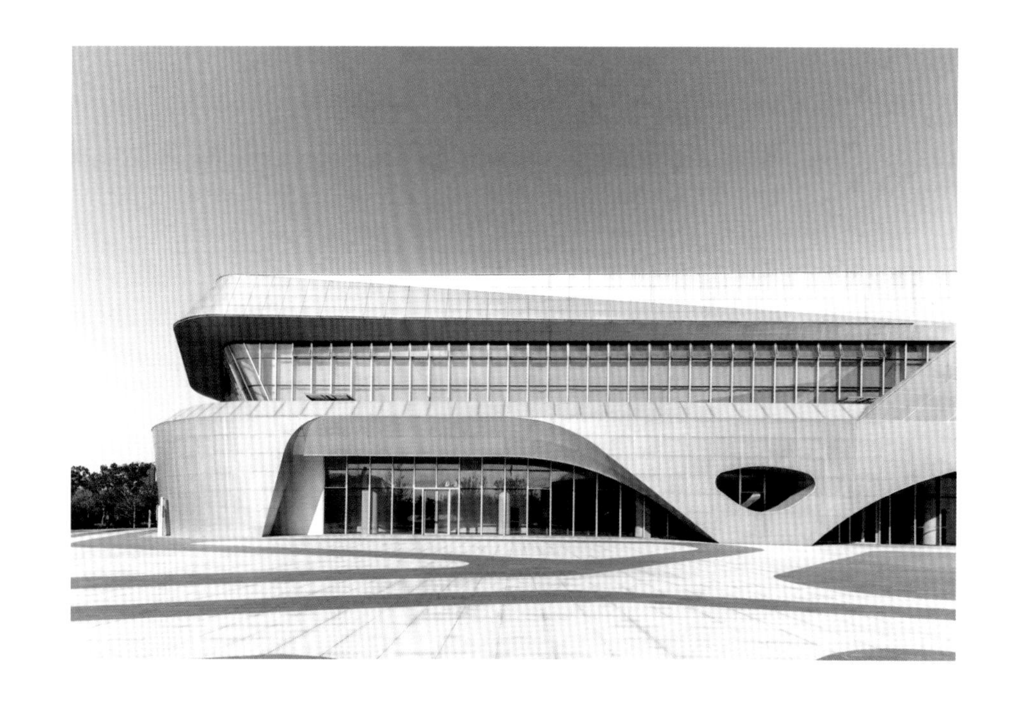

桐乡市全民健身中心
TongXiang National Fitness Center

榆阳·高新区体育运动中心
HIGH-TECH ZONE SPORTS CENTER · YUYANG

位　　置：陕西省榆林市
客　　户：北京非凡领越体育场馆运营管理有限公司
用地面积：10.21 万 m^2
建筑面积：6.17 万 m^2
功　　能：体育休闲

Location: Yulin, Shanxi Province
Client: Beijing Feifan Lingyue Stadium Operation and Management Co., Ltd.
Site Area: 102 100 m^2
Building Area: 61 700 m^2
Function: Sports & Leisure

体育纽带联结世界
SPORTS TIE THE WORLD

榆林李宁体育运动中心的形体寓意为飘带，幻化成联结世界的强劲纽带，通过文化的注入及生活方式的改变，让这座古老的城市重新焕发活力。

For the design of the Yulin Li Ning Sports Center, we adopted the shape of a floating ribbon to symbolize its transformation into a strong link to the world, helping to rejuvenate this ancient city by infusing culture and changing the lifestyles of people.

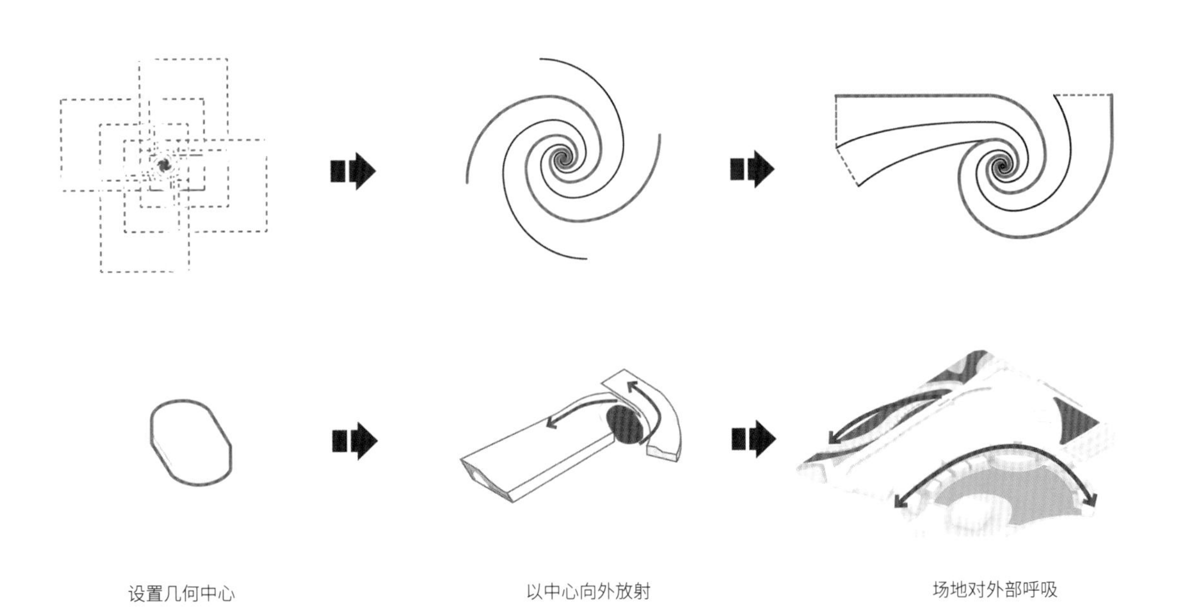
设置几何中心
以中心向外放射
场地对外部呼吸

在榆林榆阳区李宁体育小镇的设计上，我们提出了“体育 +”的规划概念，以体育为基点包罗多种规划形态，模糊空间边界的束缚，即设计团队提出的“体育共享概念”。

共享是人类未来的趋势之一。体育共享空间是我们探索的方向，用体育连接的脉络激活传统地产开发模式，复兴物理共享空间，以体育作为催化剂，与各种建筑形式产生化学效应，从而与城市达到互融、双赢。

细化到建筑上，我们将体育共享空间的理念贯彻始终，以使用者为主体思考，辅以独特的创意设计、共享的空间模式和流畅的交通动线，打造出目前国内较前沿的综合场馆（平赛一体式）体育园，并使其成为榆林城市形象的新名片。

For the development planning of Li Ning Sports Town in Yuyang District, Yulin City, the design team proposed the planning concept of "Sports+", which can be summarized as a "sports shared concept" that takes sports as the base and encompasses a variety of planning forms, blurring the boundaries of space.

Considering that sharing is one of the development trends of mankind in the future, we take the shared space for sports as the direction of exploration, aiming to activate the traditional real estate development mode by using sports as the link to revive the physical shared space, then take sports as the catalyst, to produce chemical effects with building forms, thus achieving mutual integration and win-win with the city.

We put the concept of shared space for sports into the detailed design of the building, taking the user as the main considerstion, complemented with unique creative design, shared space mode and smooth traffic flow, to create a more cutting-edge comprehensive stadium (with both daily use and competition function) in China, and make it a new business card for Yulin city image.

深圳·新世界临海揽山
MOUNTAIN BLUE · SHENZHEN

位　　置：广东省深圳市
客　　户：深圳新世界集团
用地面积：3.96 万 m^2
建筑面积：29.52 万 m^2
功　　能：住宅、公寓、商业、公共配套

Location: Shenzhen, Guangdong Province
Client: Shenzhen New World Group
Site Area: 39 600 m^2
Building Area: 295 200 m^2
Function: Residence, Apartment, Commerce, Public Supporting Facilities

自然与城市的缝合
THE SUTURE OF NATURE AND CITY

将自然环境与城市环境缝合，最大限度地利用山景资源和海景资源，实现山、城、海一体的优美宜人环境。

The original design intention of this Project is to create a beautiful and pleasant environment that integrates mountain, city and sea by stitching the natural environment with the urban space and maximizing the use of mountain and sea view resources.

SHAMA

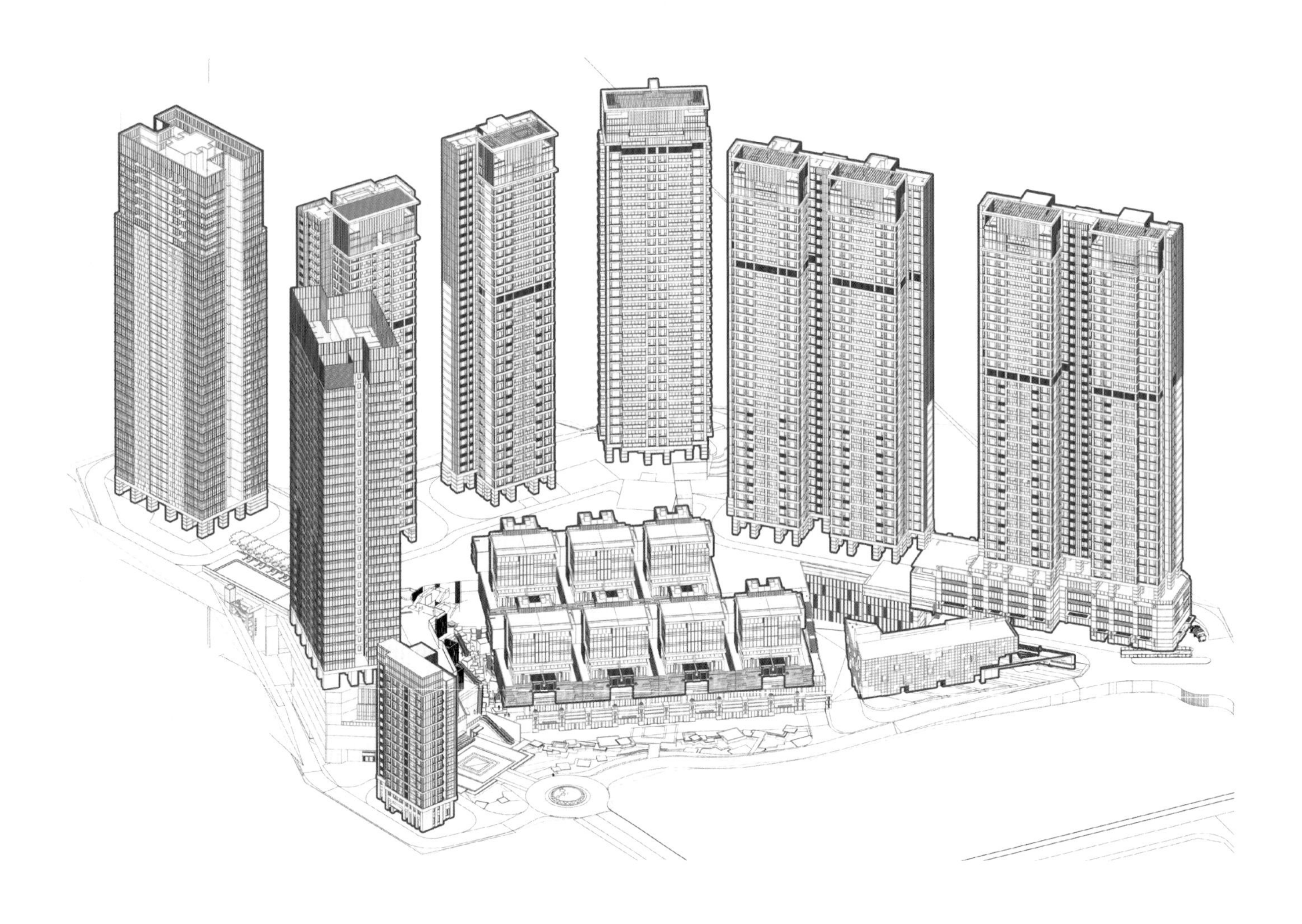

整个地块呈不规则形状。西面、北面被大南山环抱，周边建筑为多层建筑，视野开阔。在设计中，我们注重山、居住区、城市之间的关系。在设计建筑立面时我们提取了“水”元素与“山”元素形成呼应，使建筑与大南山相辅相成。立面采用线条处理，让人感觉犹如水在墙面波动，整体呈现出灵动之感。

本设计顺应山势的环抱形态，通过人工的规划还原大山的轮廓并使用一个完全自然的山坡型排列的形式，使南水建筑群和自然环境达到完美的平衡，成为豪华精品住宅小区。

The whole plot is irregular in shape, surrounded by Dananshan Mountain on the west and north, and the surrounding buildings are multi-storied with a wide view. In the design, we pay attention to the relationship between mountains, residential areas, and the city. When designing the facade of the building, we extracted the element of "water" to echo the element of "mountain", which made the building and Dananshan Mountain complement each other. By applying the line processing technique, the facade as a whole presents an agile beauty as if water is fluctuating on the wall.

Further, the design of this Project follows the encircling form of the mountain, restores the contour of the mountain through artificial planning, and uses a completely natural hillside type arrangement, so that the Nanshui building complex and the natural environment can reach a perfect balance, making this project a luxury, high-quality residential community.

郑州·电建泷悦华庭
POWERCHINA LONGYUE HUATING · ZHENGZHOU

位　　置：河南省郑州市
客　　户：电建地产 - 郑州悦宸置业有限公司
用地面积：5.99 万 m^2
建筑面积：21.17 万 m^2
功　　能：住宅、社区配套

Location: Zhengzhou, Henan Province
Client: Powerchina Realestate-Zhengzhou Yuechen Real Estate Co., Ltd.
Land Area: 59 900 m^2
Building Area: 211 700 m^2
Function: Residence, Community Supporting Facilities

承东方礼序，尊礼制大宅
ORIENT ARCHITECTURAL DESIGN WITH THE TRADITIONAL ETIQUETTE SYSTEM

设计汲取了中国传统建筑的灵动雅趣和对称秩序的精粹，以建筑作为体现中国文化底蕴的载体，构筑自然与生命和谐共存的生活体验空间。

In terms of the architectural design, we integrate the elegant and orderly design of traditional Chinese architectures. Additionally, by taking architectures as the carrier to implement traditional Chinese cultural connotations, we construct a harmonious living space for nature lives, and various life experiences.

项目整体规划呈现“一核、三轴”的格局，以两条十字景观道作为规划轴线，结合现代中式的设计语言，以一条中轴序列连接建筑群体，运用“起承转合”的手法，利用多重院落的布局形式，将整个社区环境统一成一个整体。

高低错落的形体布置体现了建筑的韵律之美，立面的柔和米色与顶部浓郁的咖啡色相映成趣，展现了建筑的和谐之道。项目通过对古典符号的提炼与演绎，唤醒社区诗意生活，重构中式人文雅居。

The overall planning of the project features the pattern of "One Core and Three Axes". Two cross-shape landscape roads are the main axes in planning, and architecture groups are arranged with a central axis sequence according to modern Chinese design characteristics. With techniques of "starting, inheriting and combining", as well as the layout form of multiple courtyards, the whole community is unified as a complete environment.

The high and low architectural layout reflects a kind of rhythmic beauty. Meanwhile, the soft beige color of the facade and the dark coffee color on the top collaboratively feature a sense of harmony. This project, by refining and deducing classical symbols, awakens the poetic life in the community and reconstructs the elegant Chinese-style residence.

深圳·华润银湖蓝山
CRL BLUE MOUNTAIN · SHENZHEN

位　　置：广东省深圳市
客　　户：华润（深圳）地产发展有限公司
用地面积：5.87 万 m^2
建筑面积：46.11 万 m^2
功　　能：住宅、公寓、商业

Location: Shenzhen, Guangdong Province
Client: China Resources Land (Shenzhen) Co., Ltd.
Site Area: 58 700 m^2
Building Area: 461 100 m^2
Function: Residence, Apartment, Commerce

不离红尘外，自在山水间
ENJOY THE BEAUTY OF THE LANDSCAPE WITHOUT BEING AWAY FROM THE HUSTLE AND BUSTLE OF THE CITY

打开的视线通廊将外部景观资源引入小区内部，创造“不离红尘外，自在山水间”的居住感受。

Through the design of open view corridor, this project introduces the external landscape resources into the interior of the community, creating the living experience of "enjoying the beauty of the landscape without being away from the hustle and bustle of the city".

宽景雅宅持销中

规划方案采用板点结合的规划形式，在中部围合成超大尺度中心花园，实现楼间距东西向最大 240 m、南北向最大 120 m 的优享空间。高低错落有致。通过数条视线通廊，将山水景观资源引入内部，形成了生态绿色人文居住社区，同时创造出举城难觅的逾 3 万 m^2 的自然山地园林，创造“不离红尘外，自在山水间”的居住感受。

通过巧妙的竖向设计，让建筑与原生地貌互成默契，人与自然和谐统一。顺应 20 m 北高南低地势变化，采用了中心庭院式设计手法，结合室外泳池、绿化、住宅首层架空，以错落的台地与水景为庭院主题，将水体、绿地、树木、铺地等交错布局，创造出一个多层次的休闲场所，给人以“出则繁华，入则幽静”的居住体验。

The planning scheme adopts a combination of slab and point building, and encloses a large-scale central garden in the middle of the building, achieving a maximum building spacing of 240m in the east-west direction and 120m in the north-south direction. The buildings are aesthetically pleasing with their harmonious heights. In terms of design, through several view corridors, landscape resources are introduced into the community, forming an ecological green humanistic living community, while creating over 30,000 square meters of natural mountain gardens that are hard to find in the city, creating a living experience of "enjoying the beauty of the landscape without being away from the hustle and bustle of the city".

Through the ingenious vertical design, this project achieves a tacit understanding between the architecture and the native landscape, and harmonious unity between man and nature. Further, the design of this project responds to the topographic change of 20 meters high in the north and low in the south, and adopts a central courtyard design approach, combining outdoor swimming pools, greenery, and residential first-floor elevation, with staggered terraces and water scenes as the courtyard theme, and staggering the layout of water bodies, green areas, trees, and paved areas to create a multi-level leisure place, thus creating a living experience that "A resident can experience urban prosperity when steps out of the community, and can feel the tranquility of the landscape when steps into the community".

武汉·中海中心
CHINA OVERSEAS CENTER · WUHAN

位　　置：湖北省武汉市
客　　户：武汉中海鼎盛房地产有限公司
用地面积：1.29 万m^2
建筑面积：10.78 万m^2
功　　能：办公、商业

Location: Wuhan, Hubei Province
Client: Wuhan Zhonghai Dingsheng Real Estate Co., Ltd.
Site Area: 12 900 m^2
Building Area: 107 800 m^2
Function: Office, Commerce

城市天际线的重塑与街道生活的缝合
RESHAPING OF CITY SKYLINE AND STITCHING OF STREET LIFE

本案通过挺拔、简洁、洗练的建筑形态记录着武汉的城市肌理，将绿色、开放的设计灵魂渗透在建筑的各个方面。

This project records the urban texture of Wuhan through the upright, simple, and concise architectural form, and the green and open design spirit is embodied in all aspects of the architecture.

Dior

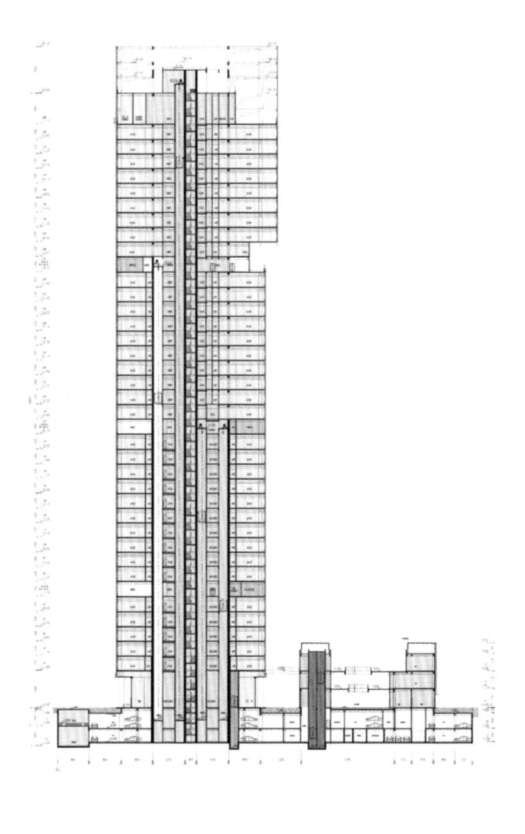

中海中心以武汉“两江三镇”的城市格局为灵感，通过对建筑形体的分解重组以及不同幕墙肌理的拼贴组合，使得建筑的各个立面和而不同，形成丰富的建筑表情，成为极具武汉地域特色的城市名片。

办公塔楼高 194.1 m，整个塔楼形态简洁明了，方正实用，内部空间合理高效。通高的玻璃幕墙以及竖向遮阳金属杆件为办公空间提供了优质的工作微环境。位于街角的多层商业表皮重复了高层体量表皮处理的建筑语汇，统一且内敛。办公楼入口大堂处低矮出挑的雨蓬及向外水平延展的城市广场，增强了建筑水平向度的延展。

Inspired by the urban pattern of "Two Rivers and Three Towns" in Wuhan, China Overseas Center makes the facades of the building harmonious but different through the decomposition and reorganization of architectural forms and the collage combination of different curtain wall textures, forming rich architectural expressions and becoming a city card with Wuhan regional characteristics.

The office tower is 194.1m high, with a simple and clear, square and practical shape, and the internal space has been used reasonably and efficiently. The through-height glass curtain wall and vertical shading metal bars provide a quality working micro-environment for the office space. The multi-storey commercial skin at the corner repeats the architectural vocabulary of the high-rise volume skin treatment in a unified and introspective manner. The low canopy at the entrance lobby of the office building and the horizontally extended city street square enhance the horizontal extension.

南京·仁恒置地广场
YANLORD LANDMARK · NANJING

位　　置：江苏省南京市
客　　户：南京仁恒江洲房地产开发有限公司
用地面积：4.51 万㎡
建筑面积：18.98 万㎡
功　　能：商业、酒店、办公

Location: Nanjing, Jiangsu Province
Client: Nanjing Yanlord Jiangzhou Real Estate Development Co., Ltd.
Site Area: 45 100 m^2
Building Area: 189 800 m^2
Function: Commerce, Hotel, Office

镶嵌在长江中的一颗绿宝石
AN EMERALD EMBEDDED IN THE YANGTZE RIVER

尊重地域环境和城市规划，以自然、生态的规划理念创造可持续发展的生态环境。

The design of this project upholds respect for the regional environment and urban planning and creates a sustainable and ecological environment with a natural and ecological planning concept.

仁恒置地广场项目位于南京长江隧道以南，江心洲·生态科技岛的中部组团。以科技为先导，以时代为坐标，强调“空间多元”、“交通立体”、“自然生态”等规划理念，建设集科技办公、风情商业、酒店公寓于一体的综合体项目。

项目建筑体量平切，退让出城市广场，以友好包容的姿态迎接从城市公园和地铁汇聚而来的大量人流，将开放街区的理念运用到整个项目，打造横穿基地的城市空间轴线。

尊重周边已建成的城市肌理，沿中新大道塑造整齐而连续的现代城市界面，同时体现绿色生态的人文关怀。

Yanlord Landmark is located in the central group of Nanjing Jiangxinzhou Ecological Science and Technology Island, south of the Yangtze River Tunnel. Guided by science and technology, this project takes the times as the coordinates and emphasizes planning concepts such as "diversified space", "three-dimensional transportation" and "natural ecology", aiming to create a complex project integrating technology-based offices, stylish commerces, and hotel apartments.

The building retreats from the city square through the volumetric cut method, and welcomes the large number of people coming from city parks and subways in a friendly and inclusive manner, applying the concept of open neighborhood to the whole project and creating an urban spatial axis across the base.

The design of the project fully respects the cempleted urban texture of the surrounding area, shaping a neat and continuous modern urban interface along Zhongxin Avenue, while reflecting the humanistic care of green ecology.

YANLORD LAND

安徽·黄山置地黎阳 in 巷
LIYANG IN LANE OF HUANGSHAN LAND · ANHUI

位　　置：安徽省黄山市
客　　户：黄山置地投资有限公司
用地面积：7.88 万㎡
建筑面积：6.25 万㎡
功　　能：商业、文旅

Location: Huangshan, Anhui Province
Client: Huangshan Land Investment Co., Ltd.
Site Area: 78 800 m^2
Building Area: 62 500 m^2
Function: Commerce, Cultural Tourism

日游黄山，夜泊黎阳
VISIT HUANGSHAN MOUNTAIN IN THE DAY AND STAY IN LIYANG AT NIGHT

设计以黎阳老街作为传承徽州文化的重要载体，采用保护、提炼、创新的设计手法，以保留黎阳老街的历史记忆，唤醒城市活力。

The design of this project takes Liyang Old Street as an important carrier to inherit Huizhou culture and adopts a design approach of preservation, refinement, and innovation to activate urban vitality while preserving the historical memory of Liyang Old Street.

在保护原老街蜿蜒曲折的肌理及修缮有保留价值的传统民居为原则的基础上，规划形态取自徽派村落自由生长的肌理和神韵。建筑单体平面设计保持传统徽州民宅内部空间“天井明堂”、“四水归堂”、“纵深序列”等基本构成方式，以明代徽州建筑风格为特色，利用现代设计手法在建筑的尺度及设施等方面保护更新。立面设计构成元素取材于传统的建筑装饰符号，营造出浓重的历史气息，唤起人们对黎阳老街的场所记忆。

On the basis of the principle of protecting the winding texture of the old streets and repairing the old dwellings with preserved value, the planning form is taken from the planning texture and charm of the free growth of Huizhou villages. The single architectural graphic design maintains the basic composition modes of the traditional Huizhou residential interior spaces such as "patio and hall", "nourishing water merges into the hall from all directions" and "depth sequence", and is characterized by Huizhou architectural style in Ming Dynasty, while using modern design techniques to protect and update the architectural scale and facilities. The facade design elements are taken from the refined traditional architectural decorative symbols to create a strong historical atmosphere to evoke memory of the place of Liyang Old Street.

云南·梦幻腾冲国际温泉度假小镇

FANTASTIC TENGCHONG INTERNATIONAL HOT SPRING RESORT TOWN · YUNNAN

位　　置：云南省腾冲市
客　　户：腾冲东云实业发展有限公司
用地面积：48.5 万m²
建筑面积：52.9 万m²
功　　能：文旅、商业、住宅

Location: Tengchong, Yunnan Province
Client: Tengchong Dongyun Industrial Development Co., Ltd.
Site Area: 485 000 m^2
Building Area: 529 000 m^2
Function: Cultural Tourism, Commerce, Residence

一场森林奇幻之旅

A MAGICAL JOURNEY THROUGH THE FOREST

以丰富的互动场景设计探寻人与人之间的亲密关系，寻找人与自然和谐相处的新型居住社区模式。

This project is designed with rich interactive scenes to explore the intimate relationship between the people and find a new model of living community where people and nature live in harmony.

展示中心基于“与生态共融”的理念，在设计中弱化了建筑与环境的边界，以六边形的几何形体为母题，通过折线的穿插引导室内外空间的渗透和交融，从形体、空间、材料各个方面表达“既秀于山林又水乳交融”的设计理念。商业街区巧妙嵌入环境中，创造出人与自然直接对话的灰空间；住宅区由多层住宅和低层住宅组合构成，形成两头高中间低的态势，建筑天际线顺应地形自然过渡。

良好的生态蕴含着无穷的价值，在建设中最大限度保护了生态环境，使人与自然和谐共存并健康发展。

Based on the concept of "integration with ecology", the exhibition center weakens the boundary between the building and the environment in its design, taking the geometry of the hexagon as the master theme and guiding the infiltration and integration of indoor and outdoor spaces through the crossing of polyline volumes, expressing the design concept of being both beautiful in the mountains and the forest and perfectly harmonious from all aspects of shape, space, and materials. The commercial area is subtly embedded in the environment, creating a grey space for direct dialogue between people and nature; the residential area is composed of a combination of multi-storey and low-rise residences, forming a situation of high at the two ends and low in the middle, with the building skyline following the natural transition of the terrain.

A good ecology contains infinite values, and protecting the ecological environment to the maximum extent in construction is an effective way for the harmonious coexistence and development of humans and nature.

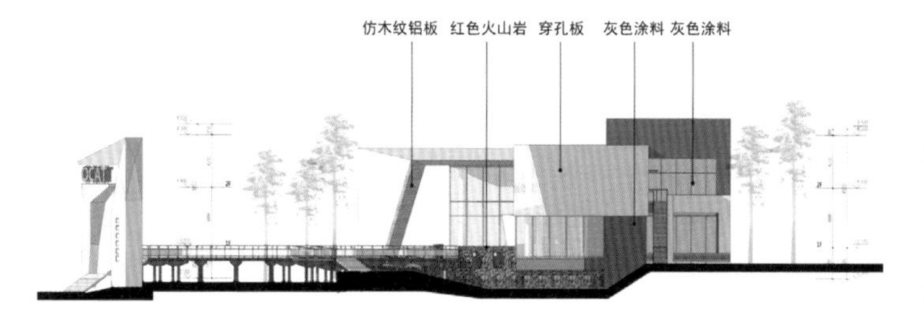
仿木纹铝板
红色火山岩
穿孔板
灰色涂料
灰色涂料

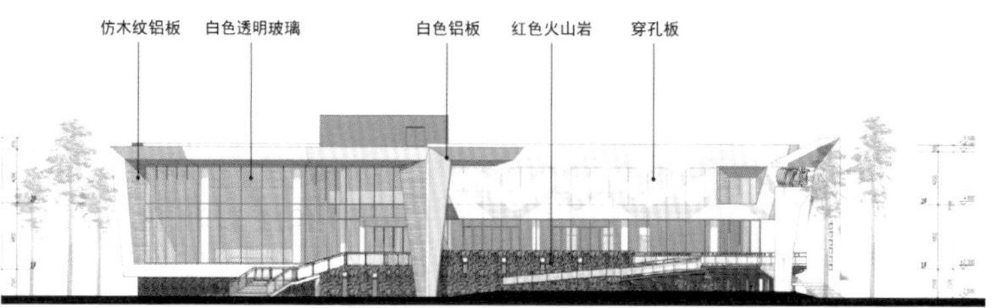
仿木纹铝板
白色透明玻璃
白色铝板
红色火山岩
穿孔板

S.T. Dupont
COSTA COFFEE
燕翔古玩店

安徽·芜湖古城一期
WUHU ANCIENT CITY (PHASE I) · ANHUI

位　　置：安徽省芜湖市
客　　户：安徽置地投资有限公司
用地面积：8.17 万㎡
建筑面积：6.93 万㎡
功　　能：文旅、商业

Location: Wuhu, Anhui Province
Client: Anhui Land Investment Co., Ltd.
Site Area: 81 700 m^2
Building Area: 69 300 m^2
Function: Cultural Tourism, Commerce

一座珍若完璧的城央古城
A PRECIOUS ANCIENT CITY IN THE CENTER OF WUHU

设计通过保护原真性历史文化资源，延续历史街道网络和肌理，以现代的规划理念，新式的建筑风格，让古城焕发新的活力。

By protecting the original historical and cultural resources, the design of this project continues the network and texture of historical streets, and makes the ancient city glow with new vitality with modern planning concepts and new architectural styles.

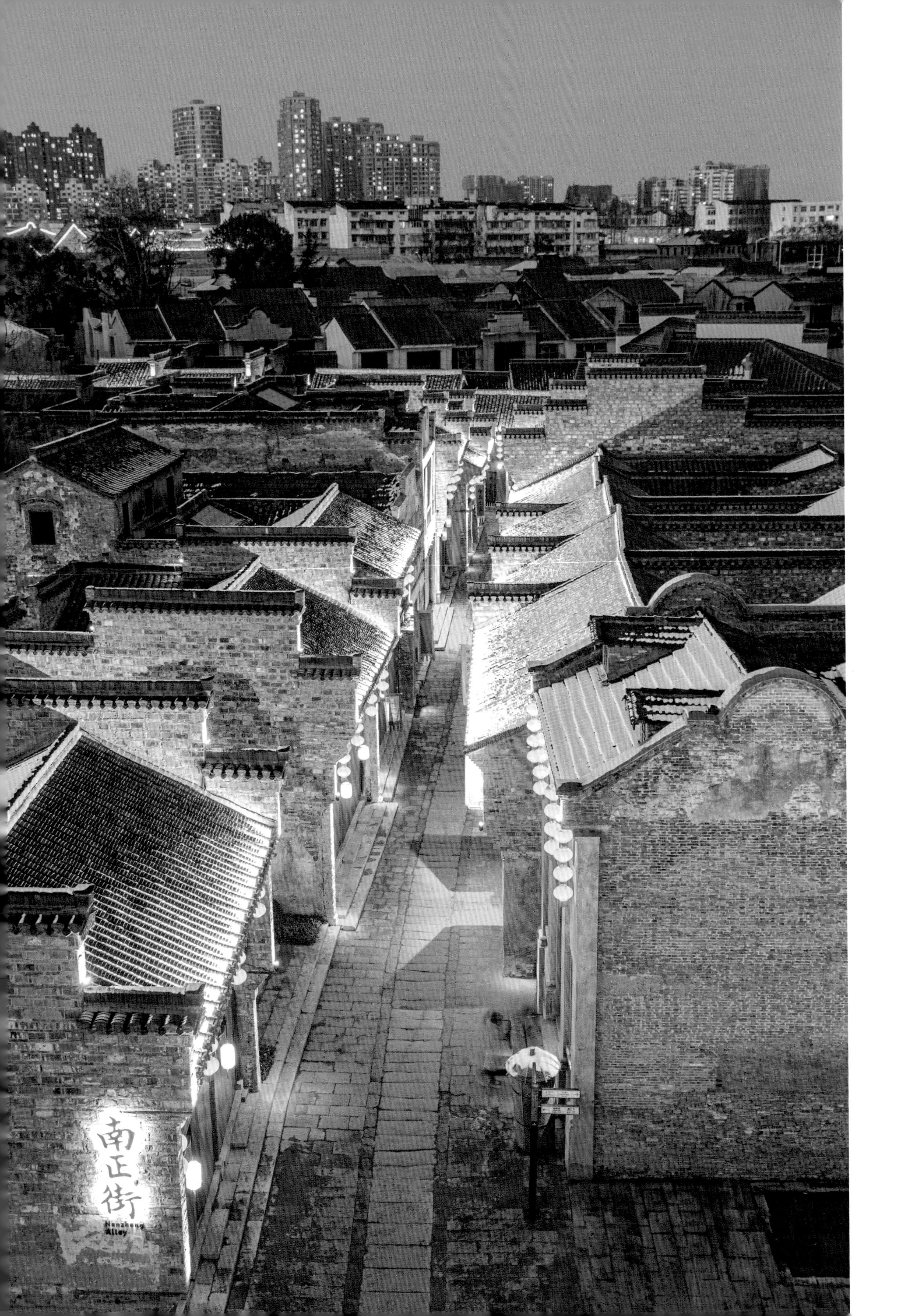
南正街
Nanzheng
Alley

保 留——保留、修缮有历史价值的构件及建筑实体，重塑古城的空间形态及规划肌理，唤醒人们对古城的历史记忆。

编 织——建筑平面设计在“天井明堂”“四水归堂”“纵深序列”“一至三进院落”等徽派建筑传统的构成方式基础上，融入“徽风西韵”的风格特征，依据现代生活需要在风格及功能方面加以改进。

再 生——提炼徽派建筑文化内涵，用现代建筑的建造方式进行演绎，使历史文化符号、传统建筑语言与现代建筑空间相融合。

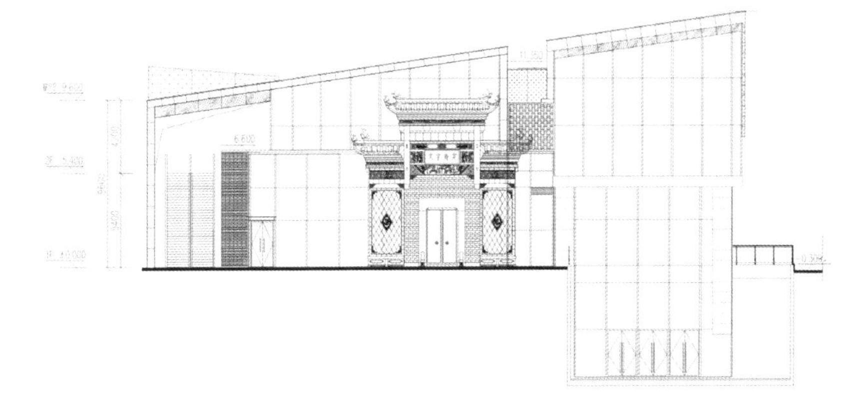

Preservation - Preserve and repair components and architectural entities with historical value, reshape the spatial form and planning texture of the ancient city to awaken people's historical memory of the ancient city.

Weaving - On the basis of maintaining the basic composition modes of the internal space of traditional Huizhou architecture, such as "patio and hall", "nourishing water merges into the hall from all directions", "deep sequence" and "a courtyard with one to three halls", the graphic design of the building incorporates the style characteristics of "Huizhou style and western rhyme", and improves the style and function according to the needs of modern life.

Regeneration - To refine the cultural connotation of Huizhou architecture, and to interpret it with the construction techniques of modern architecture, so as to integrate historical and cultural symbols, traditional architectural language, and modern architectural space.

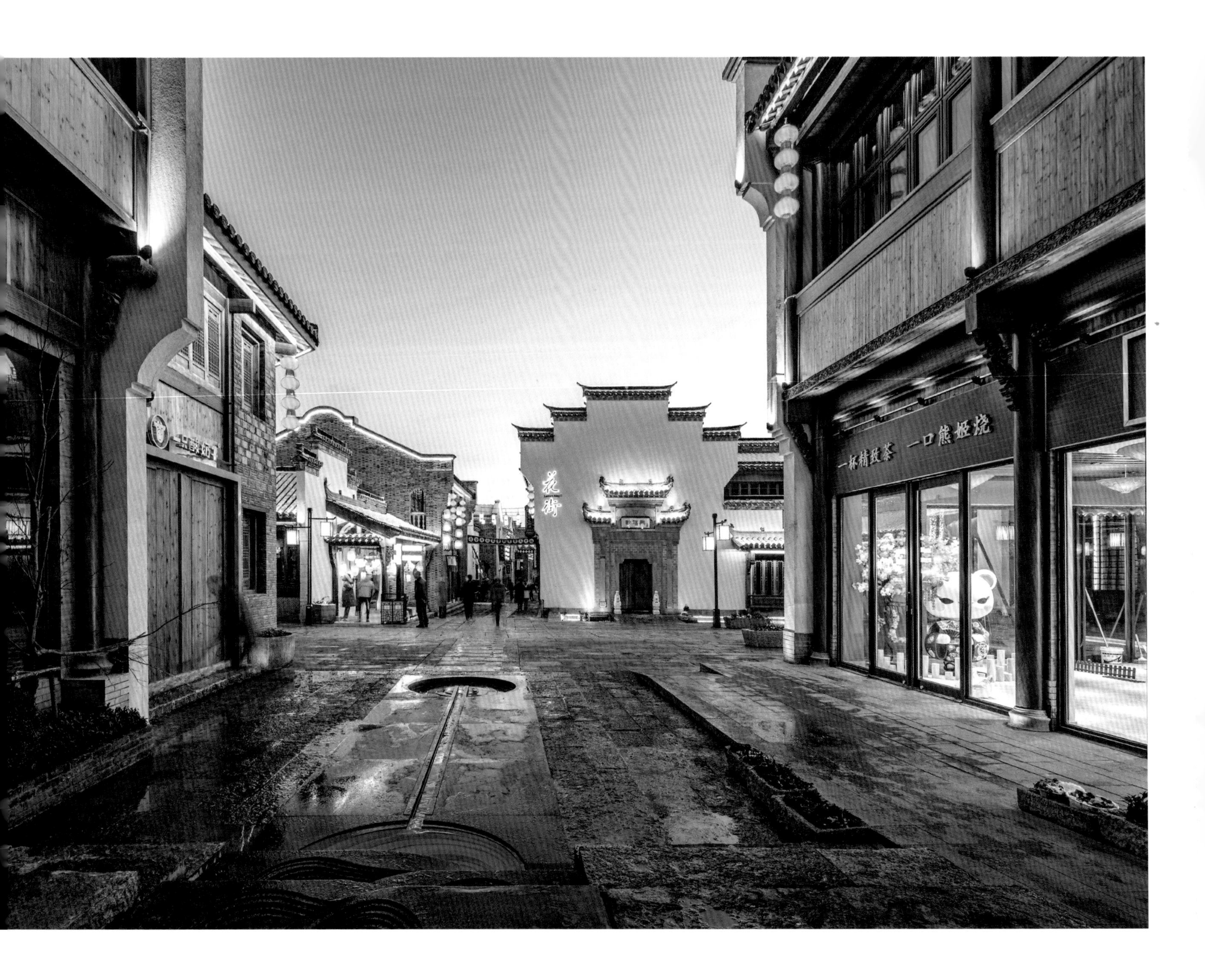
花街
一杯精致茶　一口熊姬烧

南正街

宁波·华侨城欢乐滨海
OCT JOYFUL MARINA · NINGBO

位　　置：浙江省宁波市
客　　户：宁波滨海华侨城投资发展有限公司
用地面积：19.61 万㎡
建筑面积：26.63 万㎡
功　　能：文旅、商业

Location: Ningbo, Zhejiang Province
Client: Ningbo Binhai Overseas Chinese Town Investment Development Co., Ltd.
Site Area: 196 100 m^2
Building Area: 266 300 m^2
Function: Cultural Tourism, Commerce

竞藏山海，奢享蓝海金滩
A BOUTIQUE HOME WITH A VIEW OF THE BLUE OCEAN AND GOLDEN BEACHES

建筑风格以现代、活力为主题，根据现代的商业建筑特点结合游乐项目的需求，打造出简洁、多元、个性化的商业综合体。
This project is designed in a modern and energetic architectural style. Based on the characteristics of modern commercial architecture and combined with the needs of amusement projects, this project aims to create a simple, diversified and personalized commercial complex.

宁波湾滨海旅游休闲区由于其狭长形半封闭的地块特征，形成一个避风性能良好的天然风情海港。

建筑设计以现代的设计手法打造出各具功能的三条街，分别是游乐主题街、休闲主题街、特色零售商业街。

商业区以游乐和休闲为主题功能，建筑通过坡屋顶设计及彩色线条应用打造设计特色。空间设计将游乐、商业及风情休闲商业相结合，连通南北两侧区域，打造出一条集娱乐、休闲为一体的活力商业街区。

Ningbo Bay Coastal Tourism and Leisure Area has formed a natural seaport with excellent wind shelter performance due to its narrow and semi-closed plot characteristics.

This project is designed with modern design techniques to create three streets with different functions, which are Amusement Theme Street, Leisure Theme Street, and Special Retail Commercial Street.

The commercial area takes amusement and leisure as its theme function, and the building creates its design features through a sloping roof design and colorful line application. The design of space combines amusement, commerce, and style leisure commerce, connecting the north and south sides of the area to create a vibrant commercial block that integrates entertainment and leisure.

惠州·华润小径湾

CR LAND XIAOJINGWAN · HUIZHOU

位　　置：广东省惠州市
客　　户：华润（深圳）有限公司
用地面积：17 万㎡
建筑面积：30.4 万㎡
功　　能：住宅

Location: Huizhou, Guangdong Province
Client: China Resources Land (Shenzhen) Co., Ltd.
Site Area: 170 000 m^2
Building Area: 304 000 m^2
Function: Residenec

天地合一的海边秘境

A SEASIDE SECRET WITH A HARMONIOUS LANDSCAPE

高层与多层建筑的组合，形成波浪形沿海建筑天际线，打造丰富的空间层次。

This project is designed as a combination of high-rise and multi-storey buildings, forming a wave-shaped coastal building skyline and creating a duality of rich spatial levels.

项目南临广东惠州小径湾海域，海岸线长约 2 km，并拥有大片天然优质海滩。总平面规划充分利用一线海景资源，将五星级酒店、商业服务及高低结合的住宅组团沿海岸布置。内部住宅组团尽量利用小溪形成的两岸景观，别墅区则灵活散布在山地之上。

一期住宅力争最大限度地获取海景资源。由塔式高层和弧线板式高层组合成面向大海开放的半围合组团，使每个组团获得超大尺度的临海花园空间。组团间环环相扣，令沿海建筑群天际线呈现出波浪般的律动。

This project is adjacent to Xiaojing Bay, Huizhou, Guangdong Province in the south, with a coastline of about 2 kilometers and a large area of natural high-quality beaches. The general plan makes full use of the first-line seascape resources and arranges five-star hotels, commercial services, and high-low residential groups along the coast. The internal residential clusters make the most of the views along the banks of the natural stream, while the villas are flexibly arranged over the hills.

Phase I of this project seeks to maximize the sea view resources, and is consist of tower high-rises and curved slab high-rises to form semi-enclosed clusters open to the sea, so that each cluster has a large-scale seafront garden space, and the clusters are interlocked to give a wave-like rhythm to the skyline of the coastal complex.

无锡·华侨城运河湾
OCT CANAL BAY · WUXI

位　　置：江苏省无锡市
客　　户：无锡华侨城实业发展有限公司
用地面积：14.28 万 m^2
建筑面积：41.89 万 m^2
功　　能：住宅、商业、配套

Location: Wuxi, Jiangsu Province
Client: OCT Wuxi Industrial Development Co., Ltd.
Land Area: 142 800 m^2
Building Area: 418 900 m^2
Function: Residence, Commerce, Supporting Facilities

现代的语言诠释城市记忆和复兴
INTERPRETING URBAN MEMORY AND RENEWAL IN THE MODERN LANGUAGE

现代的语言融合传统院落的概念，让建筑成为历史和文化的载体，承载一座城市的记忆。

The modern language blends with the concept of traditional courtyard, allowing the building to become a carrier of history and culture, carrying the memories of a city.

展示区设计提取现代简洁的折线元素，以双板相扣形成有趣灵动的建筑体块关系。纯粹而简洁的形体，在错动中取得平衡。

In the design of the exhibition area, the modern and simple folding line elements are extracted to form an interesting and dynamic relationship between the building blocks with the interlocking of double panels, and the pure and simple form and structure is balanced by the well-arranged form.

高层立面从“主城”和“运河”中汲取灵感，项目整体色调与古运河的建筑文化气息相结合，以折板线条划分建筑体块，通过 不对称构图削弱建筑体量感。视觉焦点区域辅以横向白色线条，增加立面节奏韵律，呈现出简约的建筑风格。

The facade of the high-rise buildings in this project draws inspiration from the "main city" and the "canal", The overall colour palette of this project combines with the architectural and cultural atmosphere of the ancient canal, with orderly folding lines dividing the building blocks and weakening the sense of building volume through an asymmetrical composition. The visual focal points are complemented by horizontal white lines to increase the rhythm of the facade, presenting a minimalist architectural style.

南京·仁恒江湾城
YANLORD JIANGWAN CENTURY · NANJING

位　　置：江苏省南京市
客　　户：南京仁恒置业有限公司
用地面积：35.31 万㎡
建筑面积：69.12 万㎡
功　　能：住宅、商业

Location: Nanjing, Jiangsu Province
Client: Nanjing Yanlord Real Estate Co., Ltd.
Site Area: 353 100 m^2
Building Area: 691 200 m^2
Function: Residence, Commerce

延续金陵城市肌理，传承长江运河文化
SUSTAINING THE URBAN TEXTURE OF NANJING CITY AND INHERITING THE CANAL CULTURE OF THE YANGTZE RIVER

江湾城以现代人文社区为指引，以南京城市文化的特征为基石，演绎现代居住小区的文化和生活形态。

Guided by a modern humanistic community and based on the characteristics of Nanjing's urban culture, Jiangwan Century renders the culture and living form of a modern residential community.

古都南京呈现“一环、两核、三轴”的城市空间结构。在设计中根据地形特点，形成了“一点、两带、四弧”的规划理念。

建筑单体设计简洁大方，通过现代手法诠释传统的比例关系。高层立面采用全铝板外墙，与东侧会展中心、南侧青奥中心和谐统一。低层公建立面提取南京传统文化元素，延续古城文脉，隐喻文化传承。

The ancient capital of Nanjing has a spatial structure of one ring, two cores, and three axes. In the design, the planning concept of one point, two belts and four arcs has been formed according to the terrain characteristics.

The single building design is simple and generous, and the traditional proportion relationship is interpreted through modern techniques. The facade of the high-rise building is made of aluminum plate, which is in harmony with the Convention and Exhibition Center on the east side and the Youth Olympic Center on the south side. The facade of the low-rise public buildings extracts the traditional cultural elements of Nanjing, continues the cultural context of the ancient city, and metaphors the transmission and inheritance of culture.

福州·福建海峡银行
HAIXIA BANK OF FUJIAN · FUZHOU

位　　置：福建省福州市
客　　户：福建海峡银行
用地面积：0.85 万㎡
建筑面积：6.60 万㎡
功　　能：银行、办公

Location: Fuzhou, Fujian Province
Client: Haixia Bank of Fujian
Site Area: 8500 m^2
Building Area: 66 000 m^2
Function: Banking, Office

建筑即精神，赋予建筑文化使命
ARCHITECTURE IS SPIRIT, WHICH ENDOWS ARCHITECTURAL CULTURE WITH MISSION

办公塔楼由两个形体互相倚靠形成，这种共同成长的态势也隐喻银行与投资公众的互相对应关系。

The office towers are composed of two leaning forms that grow together to symbolize the mutual relationship between the Bank and the investing public.

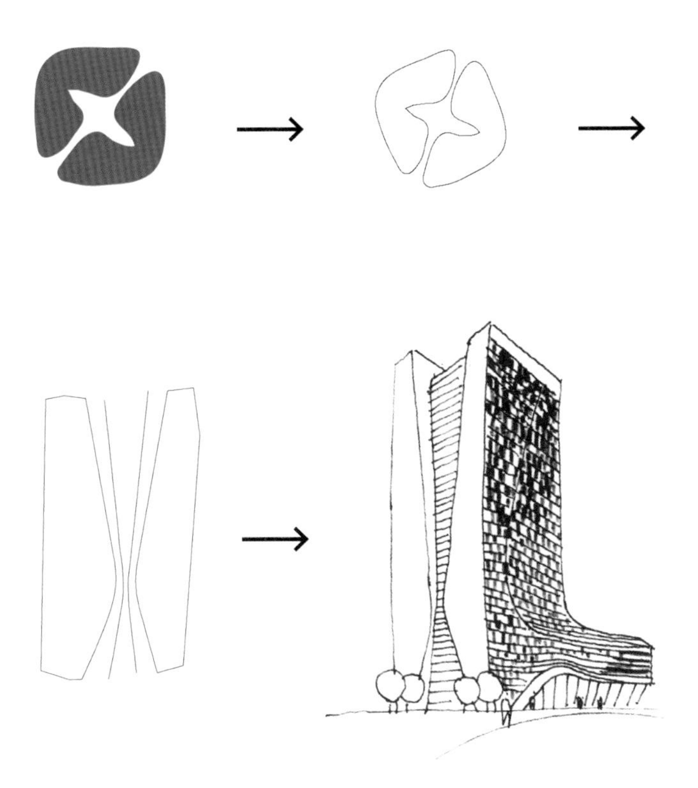

作为银行办公的建筑，设计遵循大气稳重的基本原则，实现银行建筑对公众展示的形象要求。利用原有规划和用地条件，充分尊重轴线关系，让设计能够反映与周边环境的关系，建立最基本的城市空间联系。

设计中，建筑塔楼折射着本地的历史和现状，设计将多重因素整合概括为两种积极向上的符号，这两种符号象征着银行自身与投资公众的角色及以投资需求和收益回报的关系。

从裙房开始，形体平滑扭转成构成塔楼的两个体量并向上延伸，最终构成办公塔楼的同时形成强烈攀升的视觉形象。

As a bank building, the design follows the basic principles of being dignified and steadfast, catering to the image that bank buildings present to the public. This project makes use of the original planning and site conditions, fully respecting the axial relationships to reflect the relationship with the surrounding environment and establish the most basic urban spatial connections.

In terms of design, the building tower reflects the history and status of the local area, and the design integrates multiple factors into two positive dynamics, which symbolize the role of the Bank itself and the investing public, and the relationship between investment needs and return on investment.

Starting from the podium, the body of the building smoothly twists into two volumes that constitute the tower and extends upwards, eventually forming the office tower while creating a strong visual image of the ascent.

深圳 · 中洲 πMall
ZHONGZHOU π MALL · SHENZHEN

位　　置：广东省深圳市
客　　户：深圳市中洲宝城置业有限公司
用地面积：1.49 万 m^2
建筑面积：12.71 万m^2
功　　能：商业

Location: Shenzhen, Guangdong Province
Client: Zhongzhou Baocheng Real Estate Co., Ltd.
Site Area: 14 900 m^2
Building Area: 127 100 m^2
Function: Commerce

白，即为彩
WHITE IS THE COLOR

建筑形体起源于公园里堆叠的石头，采用连续（灵动线条）与非连续（虚空）的手法塑造抽象的石头，与灵芝公园遥相呼应。

The architectural form of this Project originates from the stacked stones in the park. Continuous (dynamic lines) and discontinuous (empty space) techniques are used to create abstract stones, which echo Lingzhi Park.

Adhering to the idea of intensive land-saving design, this project adopts a huddle-centralized composition for both commerce and apartments, and creatively practices the concept of "3 in 1" tower. And it also pays attention to the simple and complete image of the interface along the urban road, echoing Lingzhi Park and forming a new landmark in Bao'an.

The plane and interior design of the building adopt the design technique consistent with the facade elements, forming a rich space interface. The commercial part of this project consists of 4 floors above ground (5 floors locally) and 2 floors underground. The interior takes three atrium spaces as nodes, links the commercial moving lines, and aggregates the surrounding places to create a three-dimensional, comprehensive, and changing space.

The whole building presents a tall and straight volume and the design of horizontal lines realizes the effect of organic circulation, highlighting the smart and simple image of the building. The architecture is highly matched with the large-scale cantilever, which has shaped the landmark buildings in this district.

项目的设计思想为集约化节地型设计，商业与公寓均采用抱团集中式构图，创造性尝试"3in1"三合体塔楼概念，注重沿城市道路界面的简洁、完整形象，与灵芝公园相呼应，形成宝安新地标。

建筑平面及室内设计采用与立面元素一致的设计手法，形成丰富的空间界面。项目商业部分为地上 4 层（局部 5 层），地下 2 层。内部以 3 个中庭空间为节点，将商业动线组织起来，聚合周边场所，创造出一个立体、综合、变化的空间。

建筑整体以挺拔的体量，加以横向线条的有机流转，强调建筑灵动、简洁形象。建筑高度配合大尺度的悬挑，塑造出了本地区的标志性建筑。

泰山·东部游客集散中心
TOURIST DISTRIBUTION EAST CENTER · MOUNT TAI

位　　置：山东省泰安市
客　　户：泰安绿漫地产有限公司
用地面积：8.67 万 m^2
建筑面积：0.90 万 m^2
功　　能：展览、商业

Location: Tai'an, Shandong Province
Client: Tai'an Lvman Real Estate Co., Ltd.
Site Area: 86 700 m^2
Building Area: 9000 m^2
Function: Exhibition, Commerce

一石成山，文化与活力的糅合
A STONE LIKE A MOUNTAIN - A COMBINATION OF CULTURE AND VITALITY

从山石中汲取创作灵感，以多维度的理念构筑起富有力量、苍茫高耸的泰山场所精神。

This project draws creative inspiration from stones to build a powerful, boundless, and towering place spirit of Mount Tai with multi-dimensional concept.

The design is based on the image of Mount Tai and Mount Tai Stone throughout this project, forming a stable, novel and amazing architectural image, while the building's upright, strong and coherent posture subtly responds to the rich natural and cultural resources in the surrounding area, resolving the abruptness of the landscape environment brought by the Distribution Center, and effectively connecting the public space with multiple characteristics of Mount Tai.

The "folded image of multiple mountains and undulating earth" is constructed with "the first layer of mountains" as the real Mount Tai in the distance, "the second layer of mountains" as the mountain-shaped building volume, "the third layer of mountains" as the facade of the ink mountains, and "the fourth layer of mountains" as the geological landscape, and the four layers of mountain scenery are continuous and dreamlike. The design of "multiple imageries of Mount Tai" achieves a precise response to the site culture of Mount Tai.

The clear form, dignified shape, and stable architectural image, fully satisfies the functional requirements while interpreting the unique humanistic style of Mount Tai landscape.

泰山与泰山石意象贯穿项目始终，整体形成稳如泰山、石破天惊的建筑形象。挺拔、凌厉、连贯的形体姿态，巧妙地回应了周边丰富的自然以及文化资源，化解集散中心给景观环境带来的突兀感，又顺势将泰山多种特质公共空间进行有效连接。

“多重山与大地起伏的褶皱意象”构筑出“一重山”即远处真实的泰山，“二重山”为山型的建筑体量，“三重山”为水墨山脉的立面，“四重山”为仿若地质的地景，四重山景一连贯之，如梦似幻。“泰山的多重意境”设计实现了对泰山场所文化的精准回应。

形体明确、造型端庄、稳如泰山的建筑形象，在充分满足功能需求的同时诠释了泰山景观独特的人文风貌。

海南·国际会展中心二期
INTERNATIONAL CONVENTION & EXHIBITION CENTER (PHASE II) · HAINAN

位　　置：海南省海口市
客　　户：海南云盛投资有限公司
用地面积：17.30 万 m^2
建筑面积：19.86 万 m^2
功　　能：会展
设计单位：上行建筑设计（深圳）有限公司、深圳市柏涛蓝森国际建筑设计有限公司、深圳市柏涛环境艺术设计有限公司

Location: Haikou, Hainan Province
Client: Hainan Yunsheng Investment Co., Ltd.
Site Area: 173 000 m^2
Building Area: 198 600 m^2
Function: Exhibition
Designed by: Shenzhen H+L Architectural Design Studio
Shenzhen PT Lansn International Architecture Design Co., Ltd.
Shenzhen PT Landscape Architecture Design Co., Ltd.

展翅的海鸥
A SEAGULL SPREADING ITS WINGS

“海角的一边，屹立着一只展翅的海鸥，正向大海翱翔。”设计以海洋元素作为切入点，延续原有建筑海洋的设计概念，坚持以人为本的原则，创造一个适宜停留的公共场所。

"On one side of the cape stands a seagull spreading its wings, ready to soar to the end of the sea". The design of this project takes marine elements as the breakthrough point, continues the original design concept of the architectural ocean, and adheres to the principle of being people-oriented, creating a suitable public place to stay.

设计团队将建筑绘制成在海洋上方飞翔的海鸥，让这个面对着城市绿轴的建筑，呈现飘逸、轻盈之感。内部双曲马鞍面的交叉结构网格，注视着神秘的海洋世界。阳光投射进馆内，仿佛透过光谱仪被曲折分解的光线一般，美不胜收。

景观设计保持海岛的文化特色，以绿化带和蓝色步道为基底，打造出一个具有海洋主题特色的园林景观。结构设计将会展建筑复杂的结构空间化繁为简，将会议中心托架的跨度和长度进行精确缩减，释放出更大的建筑空间。

海南国际会展中心二期的造型自由而舒展，与一期场馆相互协调，与城市空间相互融合，与当地文化相辅相成。

The shape of the building is designed as a seagull flying over the sea, which made the building face the green axis of Haikou and present a sense of elegance and lightness. The cross structure grid of the inner hyperbolic saddle surface reveals the mysterious sea world. When the sunlight falls into the Center, it is as if the light through the spectrum is analyzed in a zigzagging way, which is beautiful.

The landscape design maintains the cultural characteristics of the island, and takes the green belt and blue trail as the base to create a garden landscape with marine theme characteristics. In terms of structural design, the design team simplifies the complex structural space of the Center by precisely reducing its span and length to release more building space.

Hainan International Convention & Exhibition Center (Phase II) presents a free and stretching form, which is in harmony with Phase I, integrates with the urban space and complements the local culture.

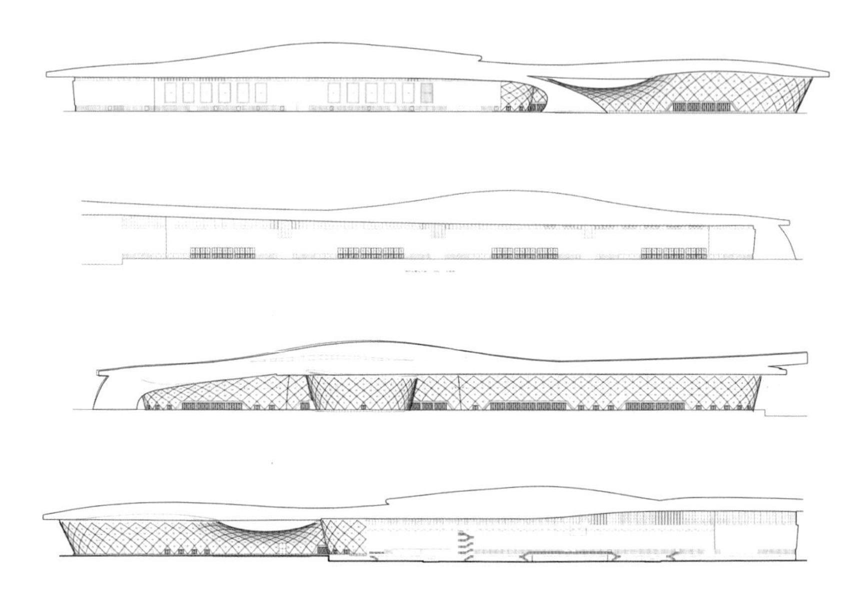

长沙·国际会议中心
INTERNATIONAL CONFERENCE CENTER · CHANGSHA

位　　置：湖南省长沙市
客　　户：长沙光达会展开发有限公司
用地面积：21.27 万 m^2
建筑面积：16.95 万 m^2
功　　能：国际会议中心
设计单位：华南理工大学建筑设计研究院（何镜堂院士工作室）、深圳市柏涛蓝森国际建筑设计有限公司

Location: Changsha, Hunan Province
Client: Changsha Guangda Exhibition Development and Operation Co., Ltd.
Site Area: 212 700 m^2
Building Area: 169 500 m^2
Function: International Conference Center
Designed by: Architectural Design &Research Institute of SCUT Co., Ltd. （He Jingtang Studio）
Shenzhen PT Lansn International Architecture Design Co., Ltd.

山水有境、天人合一
AN UNOBTRUSIVE DESIGN - AN ORGANIC INTEGRATION WITH THE NATURAL LANDSCAPE

建筑基座厚实稳重，主体轻盈飘逸，是谓虚实相生，动静相怡。

The base of the building is compacted and stable, while the main body is light and elegant, which truly interprets the coexistence of virtual and real, and the movement and static.

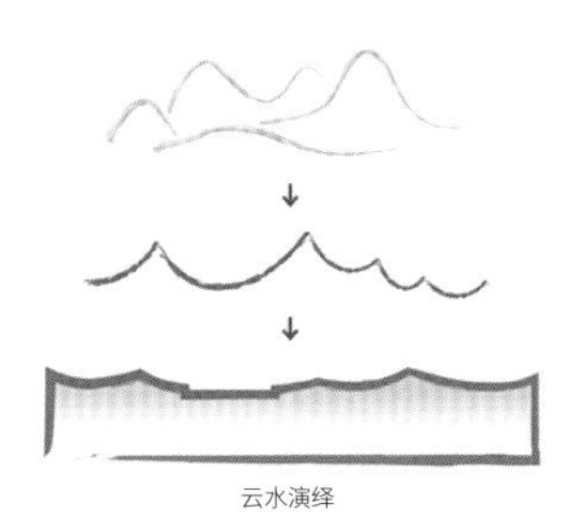
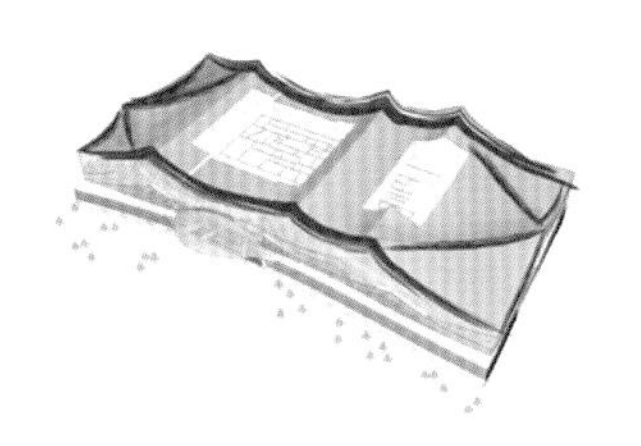

云水演绎

“山水洲城，潇湘洙泗”，设计传承长沙千古城市文化底蕴，以建筑语言奏响城市新韵律，从而打造面向世界，面向未来的城市名片。

灵动浪漫的荆楚文化，融合当地民居的抽象轮廓，实现了现代、大气且轻盈的立面形象。建筑立面分为基座与主体两部分，基座厚实稳重，主体轻盈飘逸，整体透露着灵动的气韵。

在结构设计上，进行了小、中、大震性能的设计与复核，同时对夹层对抗震性能的影响、大跨结构楼盖的舒适性、穿层柱屈曲、节点有限元等专项进行分析，确保建筑结构的安全性及空间使用的舒适性。

Located in the city of Changsha, which has the beautiful Yuelu Mountain, the majestic Xiangjiang River, the Orange Island carrying historical memories, and the Yuelu Academy famous for running schools and spreading academic culture, this project inherits the cultural heritage of Changsha for thousands of years in terms of design

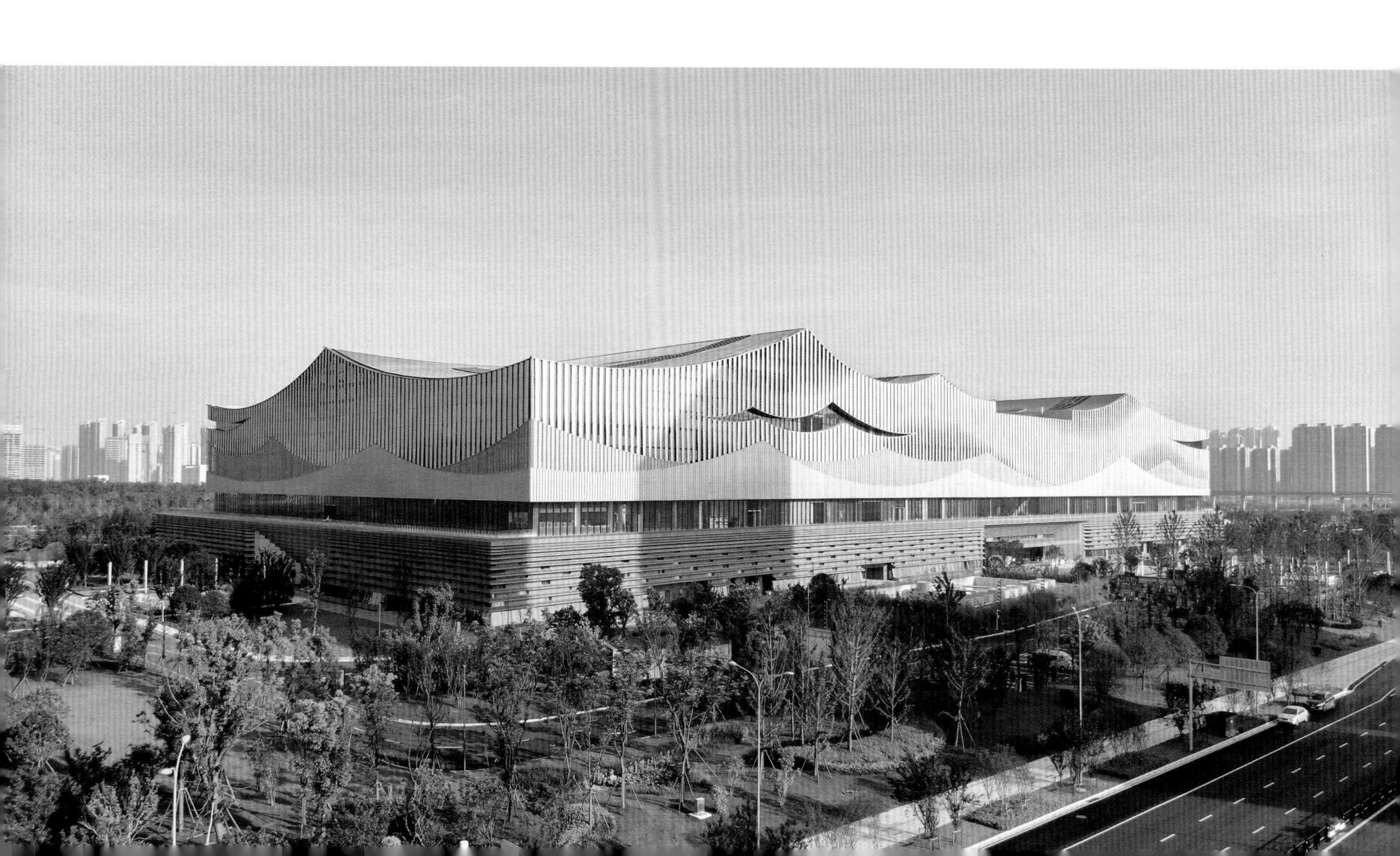

and plays a new rhythm of the city with architectural language to create a world-oriented and future-oriented city card.

The modern, atmospheric, and light facade image is achieved by integrating the abstract outline of local dwellings with the dynamic and romantic Jingchu culture. The building is divided into two parts: the base is compacted and stable, while the main body is light and elegant, revealing an overall dynamic charm.

In terms of structure, the design and review of small, medium, and large seismic performance are carried out for this project, while special analyses are conducted for the effect of the mezzanine on seismic performance, the comfort of large-span structural floor coverings, through-story column buckling, nodal finite elements, etc. to ensure the safety of building structure and comfort of space use.

榆林·榆阳中医医院
YUYANG TRADITIONAL CHINESE MEDICINE HOSPITAL · YULIN

位　　置：陕西省榆林市
客　　户：榆阳高新区管委会
用地面积：6.28 万 m^2
建筑面积：7.89 万 m^2
功　　能：综合医院

Location: Yulin, Shanxi Province
Client: Management Committee of Yuyang High-tech Zone
Site Area: 62 800 m^2
Building Area: 78 900 m^2
Function: General Hospital

开创性打造“医＋养”合一的园林式综合医院
A PIONEERING "MEDICAL + HEALTHCARE" GARDEN-STYLE GENERAL HOSPITAL

尊重塞北乡土建筑风格，结合中医药文化，采用“一心两轴四片区，三境递进八景演绎”的规划手法，营造园林式的中医医院。
This project respects the vernacular architectural style of the north of the Great Wall, combines the culture of Chinese medicine, and adopts the planning technique of "four areas with one center and two axes, and eight scenes in three stages" to create a garden-style traditional Chinese medicine hospital.

整体规划设计采用传统园林式建筑布局，通过建筑围合营造八景院落。以国医馆为核心，划分南北院区。院区南侧为城市形象轴，打造现代高效的门诊医技中心。北侧的景观轴院藏四季景观，营造予景疗愈的康养片区。本项目开创性地将“医 + 养”合一的中医理念，用于综合医院设计中，为患者提供更为全面、专业、高效、特色的医疗服务。

建筑造型传承塞北乡土建筑风格，并用现代建筑材料构造传统建筑元素。如用金属屋面、陶板、石材、格栅分别构造飞檐、灰砖、黄土窑洞、木质窗花等。现代建筑技艺让建筑更为经久耐用且低碳节能。建筑、景观、室内设计均以国医文化为概念，通过匠心设计，让医院摆脱了苍白、冰冷的刻板印象，取而代之的是绿色自然、阳光温暖、便利舒适的人性化关怀。让患者置身其中，得到身心治愈的体验，是本项目贯穿始终的设计目标。

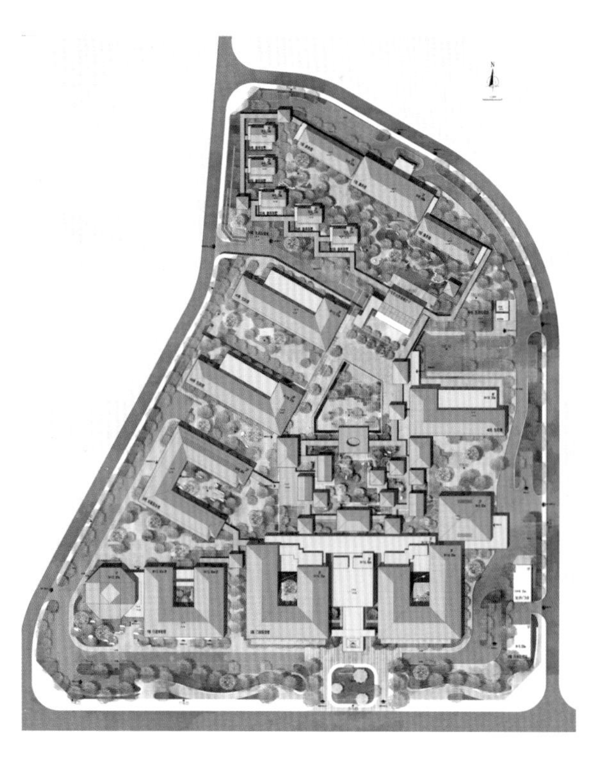

The overall planning and design of this project adopt the traditional garden-style architectural layout, creating eight scenic courtyards through the construction technique of building enclosure. With the National Medical Center as the core, it is divided into the south and north hospital zones. The south side of the hospital zone is the urban image axis, creating a modern and efficient outpatient medical technology center, while the landscape axis on the north side contains four seasons of landscape, creating a recreational area that combines scenery to heal body and mind. Further, this project pioneered the concept of "medicine + healthcare" in the design of a general hospital, providing patients with more comprehensive, professional, efficient, and unique medical services.

The building shapes in this project inherit the vernacular architectural style of the north of the Great Wall and construct traditional architectural elements with modern building materials, such as metal roofing, ceramic panels, stone, and grilles to construct flying eaves, gray bricks, loess kilns, and wooden window grilles, respectively. Modern construction techniques make the buildings more durable and ensure low carbon while achieving energy conservation. The architecture, landscape, and interior design are all based on the concept of national medical culture and are designed to reverse the stereotypical image of hospitals as pale and cold and to create the impression of a hospital as a place of green nature, sunshine, warmth, convenience and comfort, and humanistic care. It is the design goal this throughout project that patients can be healed physically and mentally by being there.

成都·中铁鹭岛艺术城
CHINA RAILWAY EGRET ART CENTER · CHENGDU

位　　置：四川成都市
客　　户：滕王阁地产、中铁二局
用地面积：4.1 万 m^2
建筑面积：26.9 万 m^2
功　　能：超高层住宅、情景文化商业

Location: Chengdu, Sichuan Province
Client: Townowner Real Estate & China Railway NO.2 Engineering Group Co., Ltd.
Site Area: 41 000 m^2
Building Area: 269 000 m^2
Function: Super High-Rise Residence, Scene Culture Commerce

老城区的文化复兴
CULTURAL RENAISSANCE OF THE OLD TOWN

现代文化和历史文化的碰撞与融合，开放混合形态社区的全新体验。
The collision and integration of modern culture and historical culture create a new experience of mixed community.

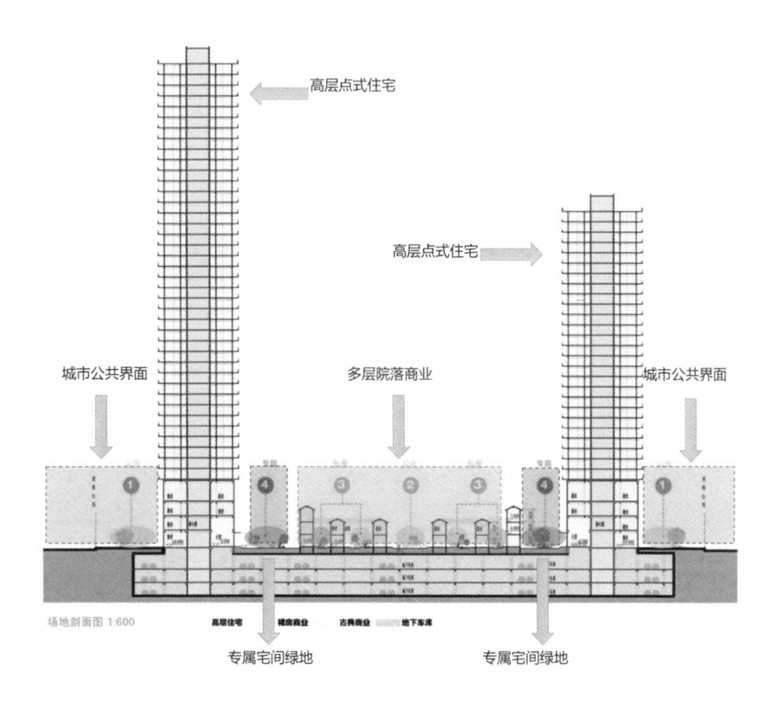
高层点式住宅
高层点式住宅
城市公共界面
多层院落商业
城市公共界面
高层住宅
裙房商业
古典商业
地下车库
专属宅间绿地
专属宅间绿地
场地剖面图 1:600

该项目是市井商业与居住区混合形态设计的一种全新尝试，立足“文化促进商业，商业带动地产，地产改变环境”的老城区复兴理念。

设计上，现代公建化的点式住宅塔楼 U 形布置，建筑形体精致明快；传统文化商业呈院落式布置于基地内部，富有人文气息。二者相互融合映衬，形成了一个完整的成都文化体验以及极致的空间体验场所，打造出一个引领都市时尚生活的新型城市中心区，让成都老城区文化在这里焕发了当代的活力。

Based on the revitalization concept of the old town that "culture promotes commerce, commerce drives real estate, and real estate changes the environment", this project is a new attempt at the hybrid design of street commerce and residential areas.

In terms of design, this project adopts a U-shaped layout of modern public housing towers, presenting a refined and bright architectural form; and traditional cultural commerce is laid out in a courtyard style inside the base, and is rich in humanistic atmosphere. The two are integrated to form a complete Chengdu cultural experience and the ultimate space experience venue, creating a new urban center area that leads the urban fashion life, allowing the culture of the old town of Chengdu to radiate contemporary vitality here.

贵阳·中铁阅山湖云著
CREG YUESHANHU YUNZHU · GUIYANG

位　　置：贵州省贵阳市
客　　户：中铁置业集团贵州有限公司
用地面积：15.17 万 m^2
建筑面积：51.58 万 m^2
功　　能：住宅、商业

Location: Guiyang, Guizhou Province
Client: China Railway Real Estate Group (Guizhou) Co., Ltd.
Site Area: 151 700 m^2
Building Area: 515 800 m^2
Function: Residence, Commerce

云水之间，为梦而著
A RESIDENTIAL COMPLEX BUILT FOR DREAMS BETWEEN CLOUDS AND WATER

在自然山水环境基调之下，以传统文人的“山水游”为空间组织线索，打造出天人合一的人居空间。

Under the tone of the natural landscape environment, this project takes the traditional literati's "landscape tour" as the spatial organization clue to create a living space where man and nature live in harmony.

造一半
造一半

空间规划错落有致，使得各楼栋的景观资源发挥到极致，以最大程度保证每栋楼都具有优质的景观面及采光面。住区内以中式园林造景为主要设计手法，寻求人居环境和自然的平衡点，实现人们对“繁华都市”“山水自然”及“清幽庭园”三层空间的完全拥有。

古典建筑与园林重新演绎形成现代典雅的风格，为居住者提供独一无二、宁静自然的情感空间；简洁明快的色彩与中轴对称的规划布局相得益彰，体现了现代典雅的大气与纯净，细腻且极具神韵。

This project is well-arranged in spatial planning, so that the landscape resources of each building are maximized to ensure that each building has a quality facade and daylighting surface, and the main design technique of Chinese gardening is used in the residential area to seek a balance between human living environment and nature, realizing the three levels of space of "prosperous city", "natural landscape" and "quiet garden".

This project reinterprets classical architecture and gardens to form a modern and elegant style, providing residents with a unique, tranquil , and natural emotional space. Further, it adopts simple and bright colors, which complement the symmetrical planning layout of the central axis, reflecting the purity and magnificence of modern elegance, which is delicate and highly divine.

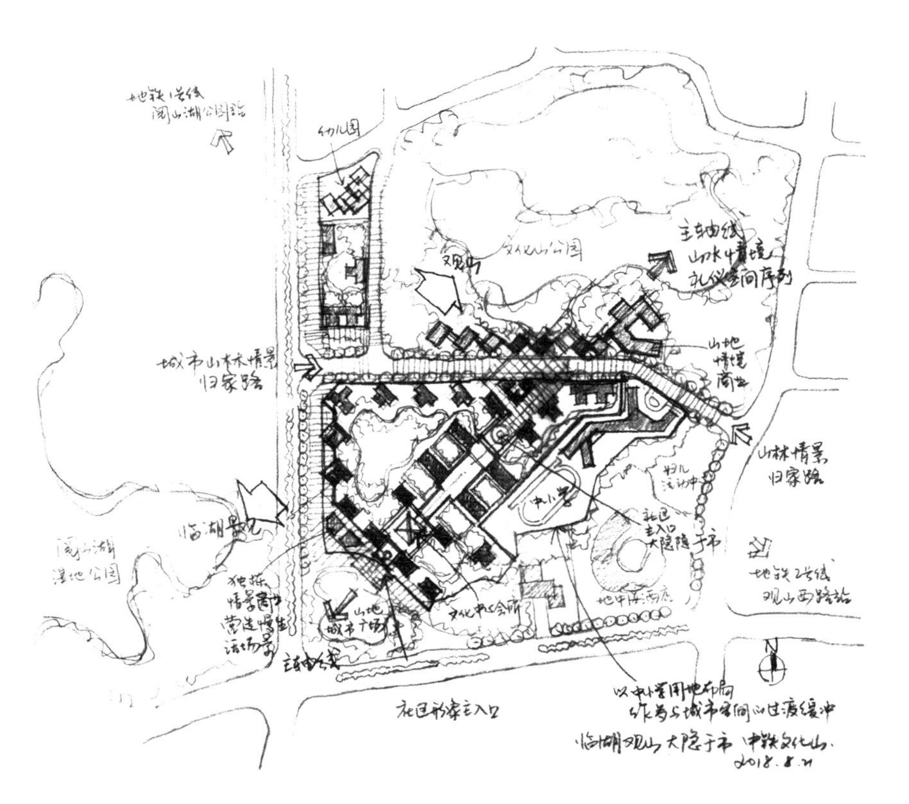
幼儿园
文化山公园
观山
主轴线
山水情境
礼仪空间序列
城市山林情景
归家路
山地
情境
商业
山林情景
归家路
中小学
社区
主入口
大隐于市
临湖景观
地铁2号线
观山西路站
地中海酒店
文化中心会所
山地
城市广场
主轴线
社区形象主入口
临湖观山 大隐于市 中铁文化山

佛山 · 粤海拾桂府
YUEHAI SHIGUIFU · FOSHAN

位　　置：广东省佛山市
客　　户：佛山粤海置地有限公司
用地面积：4.33 万 m^2
建筑面积：20.32 万 m^2
功　　能：住宅

Location: Foshan, Guangdong Province
Client: Foshan Yuehai Real Estate Co., Ltd.
Site Area: 43 300 m^2
Building Area: 203 200 m^2
Function: Residence

居者禅心
ZEN MIND OF DWELLERS

以人文脉络塑造场地精神，引山水精神注入场地记忆，化繁为简，打造精工简洁的艺术通廊。

This project aims to shape the spirit of the site with humanistic veins, to inject the spirit of the landscape into the memory of the site, to simplify the complexity and create a refined and simple art corridor.

项目引入艺术与生态的概念，通过对建筑的艺术化处理，形成与自然接洽的美学会馆。

设计上，引用岭南符号，通过立面表皮系统，串联起建筑、室内和景观。归家之门，使用有序排列的石材贴面，与自然禅意的水池交相辉映；酒店氛围式长廊，于石面与玻璃之间置入通透、轻盈的渐变穿孔铝板，巧妙搭建大虚大实的过渡，演绎出现代建筑的精致之美；水境环绕，建筑整体映入水帘，池中雕塑宛如侍者，恭迎归家的游子。

生活的哲学，几何式呈现；场地的再造，美学性还原。

This project introduces the concept of art and ecology and forms an aesthetic guild hall in contact with nature through the artistic treatment of architecture.

The Lingnan symbols are used in the design, and the facade skin system connects the building, interior, and landscape. The main entrance of the residence is designed with an orderly arrangement of stone veneer, reflecting the natural Zen pool; the promenade is designed as a hotel atmosphere promenade, with a transparent and light gradient perforated aluminum plate between the stone surface and glass, skillfully building a transition of the virtual and the real, interpreting the exquisite beauty of modern architecture; as for the water surrounding, the whole building is reflected in the water curtain, and the sculpture in the pool is like a waiter, welcoming the homecoming travelers.

This project presents the philosophy of life in a geometric way; and the re-creation of the site reflects an aesthetic restoration.

武汉·中信泰富滨江金融城阅江荟

HARBOUR CITY BUND-HIGH · WUHAN

位　　置：湖北省武汉市
客　　户：武汉泰富二零四九滨江房地产开发有限公司
用地面积：1.09 万 m^2
建筑面积：6.99 万m^2
功　　能：办公、商业

Location: Wuhan, Hubei Province
Client: Wuhan Pacific 2049 Business Development Co., Ltd.
Site Area: 10 900 m^2
Building Area: 69 900 m^2
Function: Office, Commerce

畅享滨江生态与城市繁华

ENJOY RIVERFRONT ECOLOGY AND URBAN PROSPERITY

一个与滨江生态互动，与城市生活对话的可持续城市花园；一个活力多元、绿色开放的共享社区。

This project aims to create a sustainable urban garden that interacts with the riverfront ecology and dialogues with urban life, as well as a vibrant and diverse, green and open shared community.

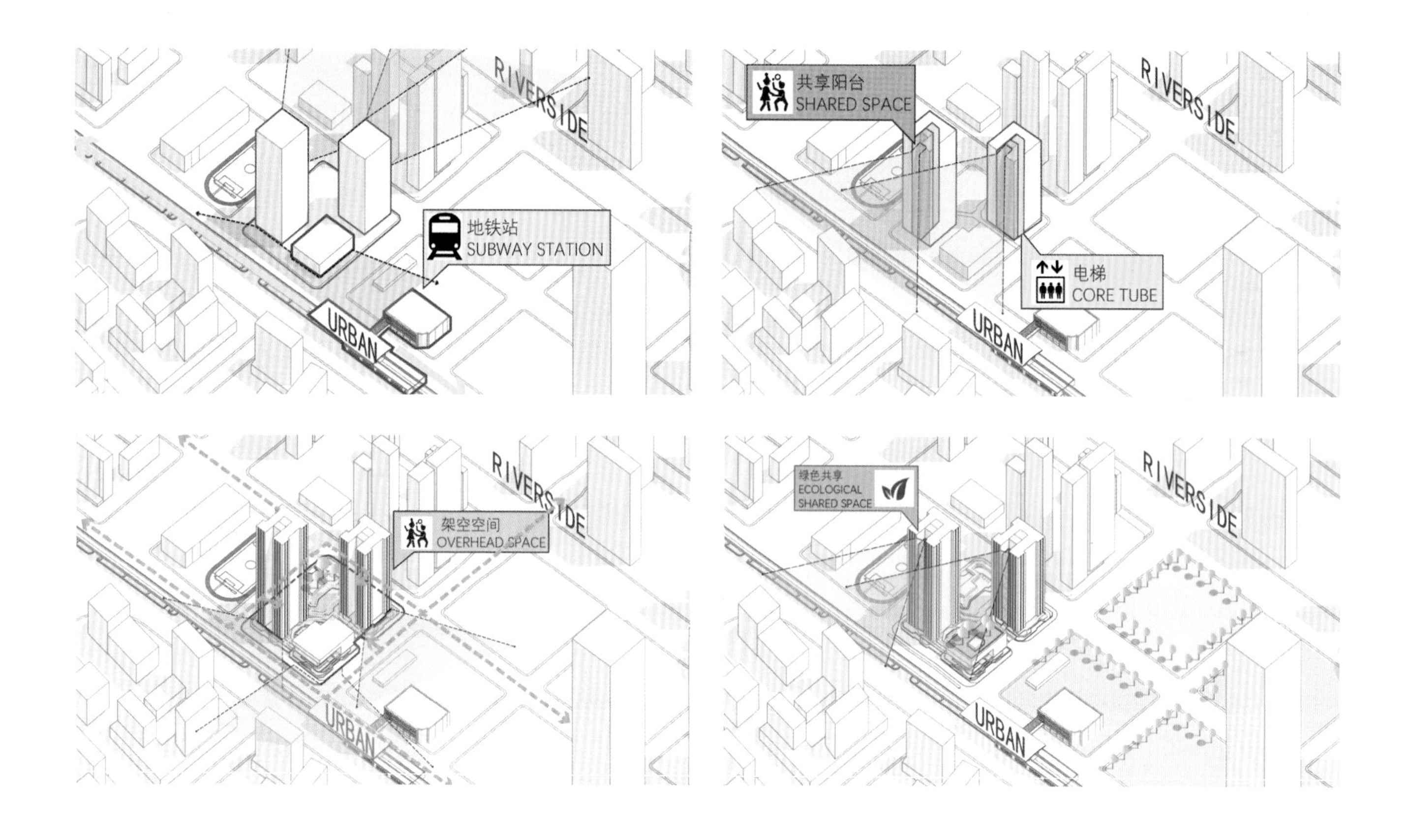
RIVERSIDE
地铁站
SUBWAY STATION
URBAN
共享阳台
SHARED SPACE
电梯
CORE TUBE
架空空间
OVERHEAD SPACE
绿色共享
ECOLOGICAL
SHARED SPACE

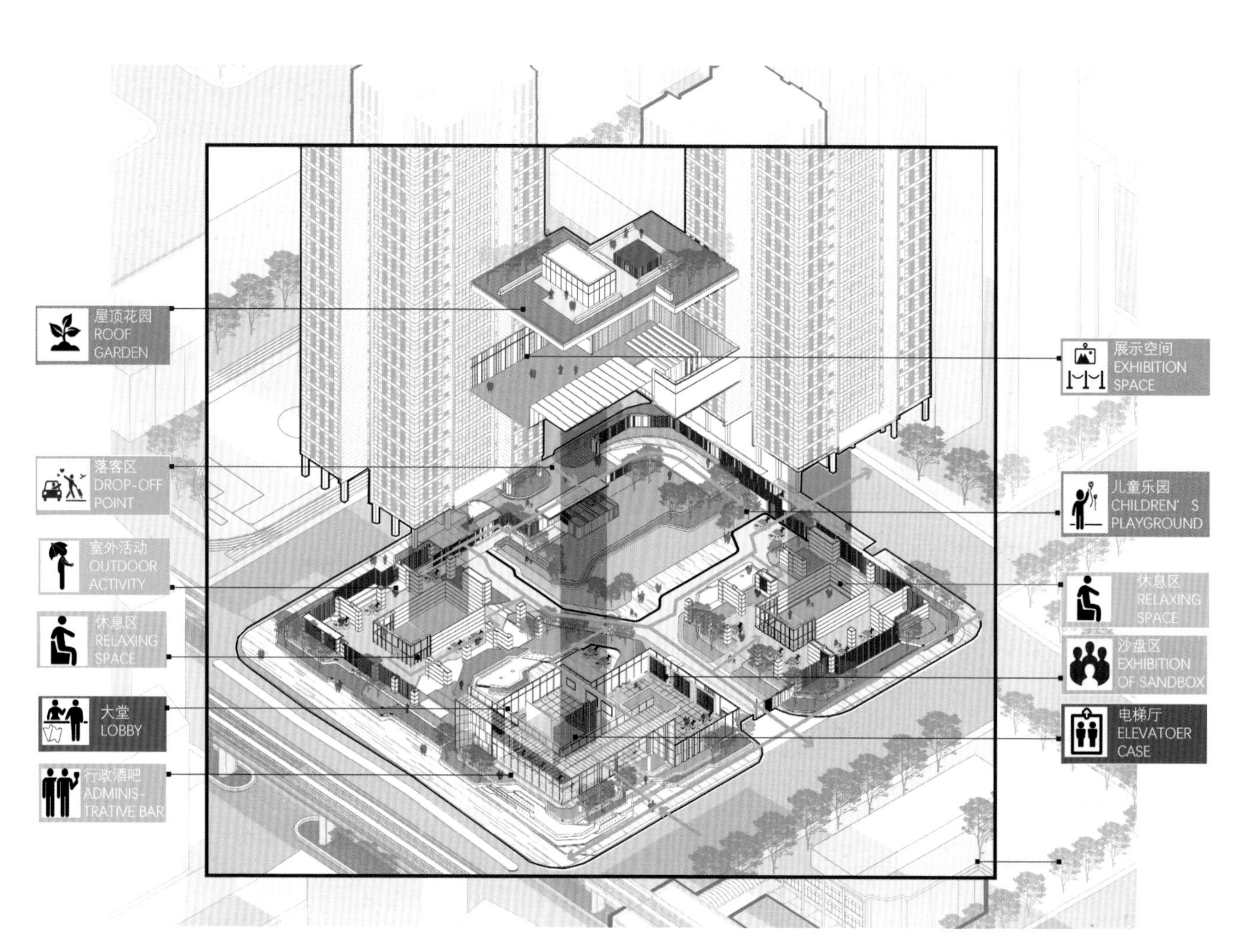
屋顶花园
ROOF
GARDEN
落客区
DROP-OFF
POINT
室外活动
OUTDOOR
ACTIVITY
休息区
RELAXING
SPACE
大堂
LOBBY
行政酒吧
ADMINIS-
TRATIVE BAR
展示空间
EXHIBITION
SPACE
儿童乐园
CHILDREN' S
PLAYGROUND
休息区
RELAXING
SPACE
沙盘区
EXHIBITION
OF SANDBOX
电梯厅
ELEVATOER
CASE

项目以“城市花园”、“标杆形象”、“生态社区”作为设计核心理念，旨在让人享受开阔视野的同时，拥有私密且明亮的空间，并与周边城市建立起紧密的关系，为都市人构建新兴的生活模式。

整体布局上，充分考虑建筑与周边的关系，将两座高低错落的塔楼以对角线的方式布局，重构出灵动的城市天际线。展示中心的设计将当地文脉融入其中，打造刚柔并济的艺术形象。塔楼的设计以简洁、现代的几何形体为基础，通过体量的凹与凸，形成灵动的体型。浅灰色纵向装饰铝板强调出建筑的垂直线条以及体量感。玻璃与铝板的虚实对比，使整个建筑形象挺拔有力，构建出城市主干道旁的标杆形象。

This project takes "urban garden", "benchmark image" and "ecological community" as the core design concepts, seeking to establish a close relationship with the surrounding city while allowing people to enjoy the view, light and private space, so as to build an emerging life mode for urbanites.

In the overall layout, the relationship between the building and the surrounding city is fully considered, and the two towers with scattered heights are laid out diagonally to reconstruct the spiritual city skyline. The design of the exhibition center incorporates the local culture to create a rigid and flexible artistic image. The design of the tower is based on a simple and modern geometric shape, which forms a spiritual body through concave and convex volume. The light gray longitudinal decorative aluminum panels emphasize the vertical lines of the building and the sense of volume, while the contrast between falsehood and reality between the glass and the aluminum panels makes the whole building image strong and upright, creating a benchmark image along the main road of the city.

5

深圳·益田益科大厦
YITIAN YIKE TOWER · SHENZHEN

位　　置：广东省深圳市
客　　户：深圳市益田世达投资管理有限公司
用地面积：1.25 万㎡
建筑面积： 10.29 万㎡
功　　能：办公、商业综合体

Location: Shenzhen, Guangdong Province
Client: Shenzhen City Yitianshida Investment Management Co., Ltd.
Site Area: 12 500 m^2
Building Area: 102 900 m^2
Function: Office, Commercial Complex

生态谷，一座会呼吸的建筑
ECO VALLEY, A BUILDING THAT CAN "BREATHE"

设计以“智慧”“生态”作为核心概念，力求建立与燕子岭公园的对话；在裙房、塔楼创造出不同层次的绿色开放空间，形成一个多维、立体的“生态谷”。

The project takes "intelligence" and "ecology" as the core design concepts, and seeks to establish a dialogue with Yanziling Park; and by creating different levels of green open space in the podium and towers, it forms a multi-dimensional and three-dimensional "Eco Valley".

50

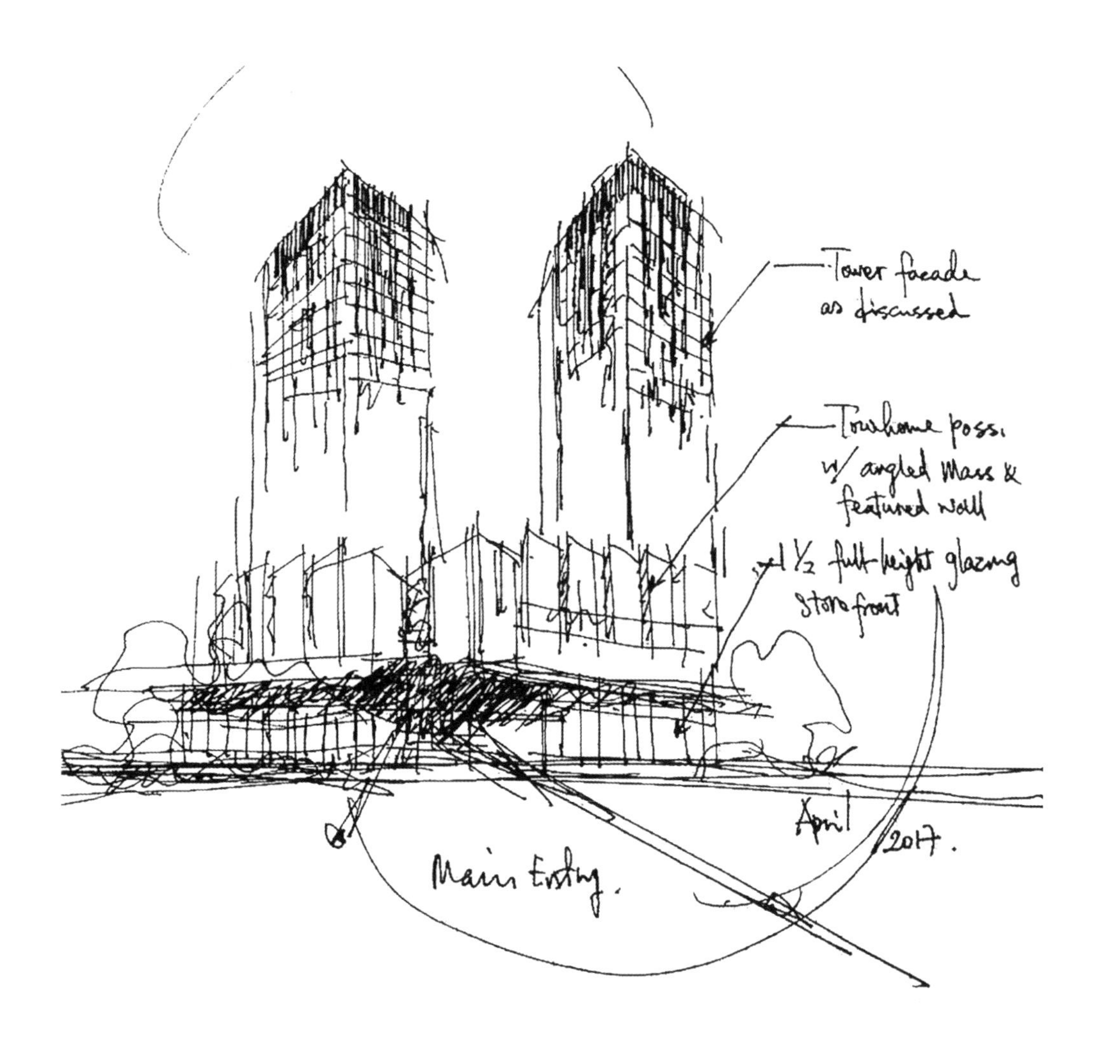

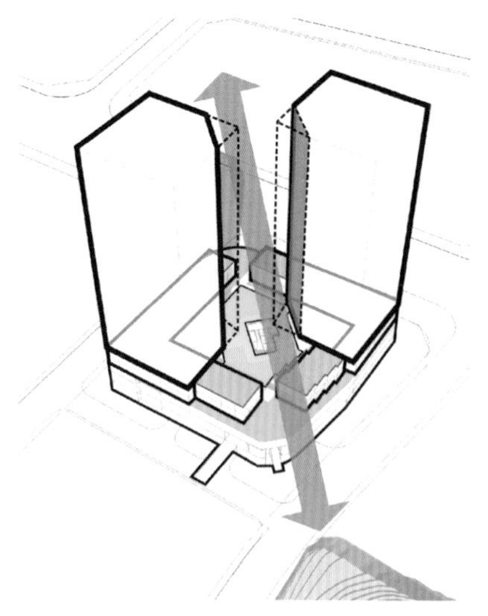

视觉通廊

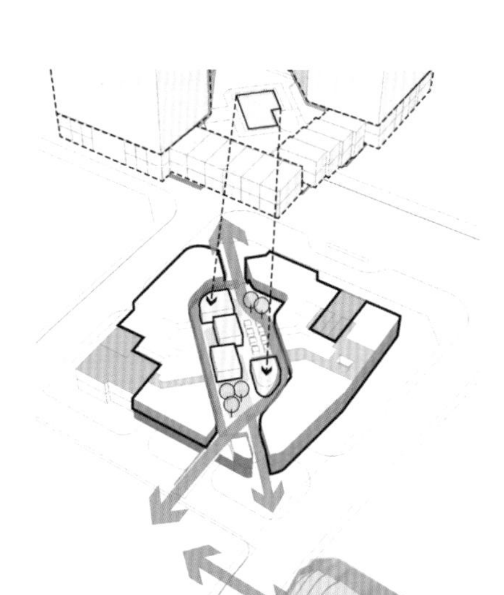

开放绿色空间

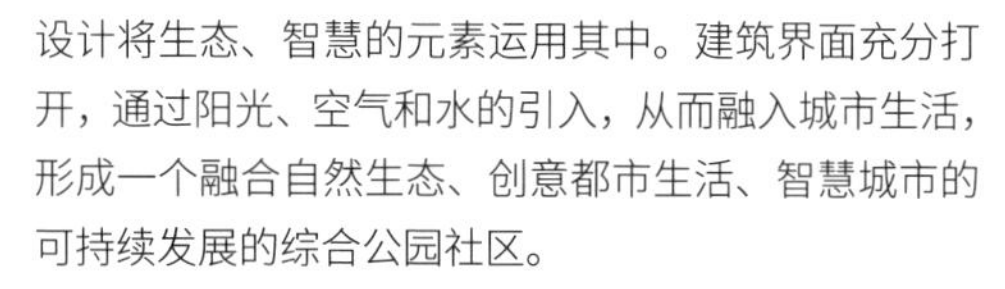

设计将生态、智慧的元素运用其中。建筑界面充分打开，通过阳光、空气和水的引入，从而融入城市生活，形成一个融合自然生态、创意都市生活、智慧城市的可持续发展的综合公园社区。

建筑布局充分顺应城市关系，塔楼沿对角布置，通过体量切割，形成视觉通廊以及公共空间轴线。立面设计将低碳、智慧的元素与之融合，塔楼采用隐形竖向条窗，外侧设穿孔铝板，在保证通风环境的同时，创造出一个更简洁、纯粹的立面形象。

Ecology and intelligence elements are the design highlights of this project. The interface of the building is fully opened, and through the introduction of sunlight, air, and water, it forms a sustainable and comprehensive park community that integrates natural ecology, creative urban life, and intelligent city.

The layout of the building fully responds to the urban relationship, with the towers arranged diagonally and cut through the volume to form visual corridors and public space axes. The facade design combines low-carbon and intelligent elements. The tower adopts invisible vertical windows and perforated aluminum plates on the outside, which not only ensures the ventilation environment, but also creates a more concise and pure facade image.

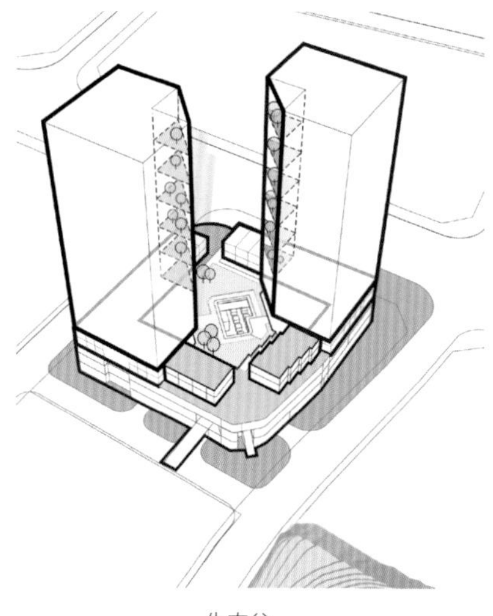

生态谷

商业空间的设计有别于传统的封闭式商场，引入“多首层”的概念，将空间充分打开，创造出一系列的广场、街巷、庭院、绿化等连续空间体验，并通过二层连廊连接燕子岭公园，实现与自然的充分互动。在各功能之间创造交流，强化与自然山体的联系。

The design of commercial space is different from the traditional closed shopping mall, the design of this project introduces the concept of "multiple ground floors" to open up the space to create a series of plazas, streets, courtyards, greenery, and other continuous spatial experiences, and connects to Yanziling Park through the corridor on the second floor, realizing full interaction with nature. In addition, this design realizes the communication between different functions and strengthens the connection with the natural mountains.

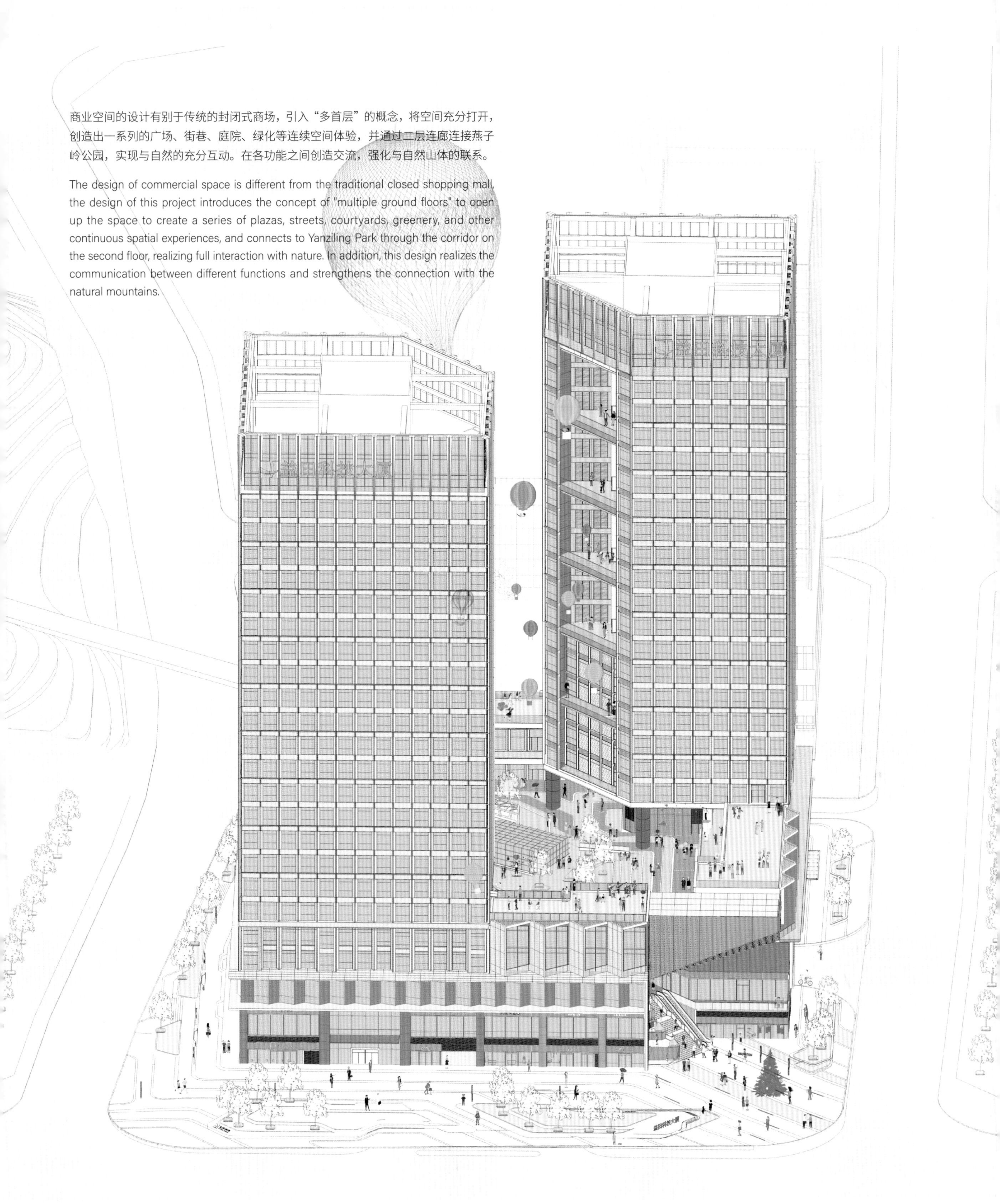

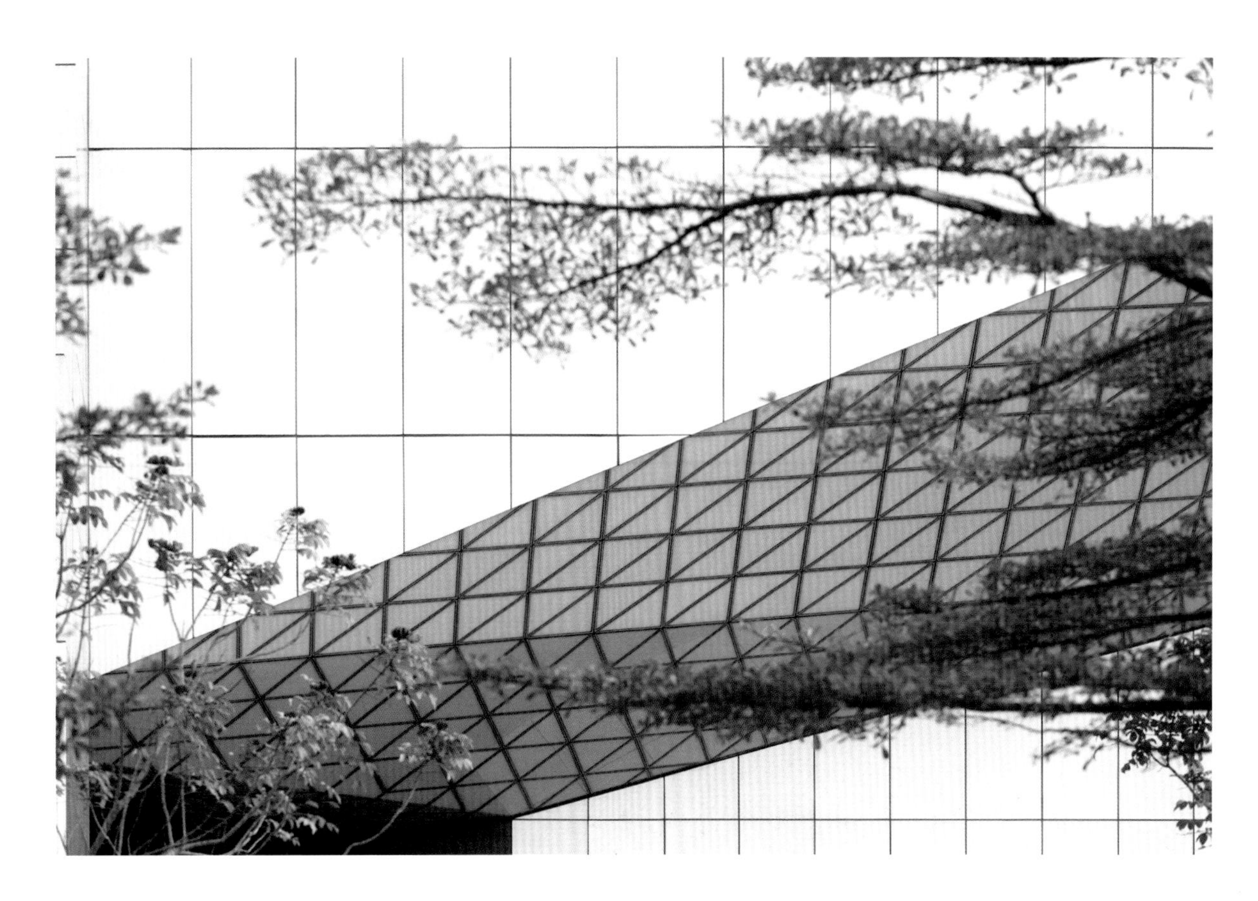

广州·南沙金茂湾
NANSHA JINMAO BAY · GUANGZHOU

位　　置：广东省广州市
客　　户：方兴地产（中国）有限公司
用地面积：4.52 万m²
建筑面积：10.02 万m²
功　　能：办公、商业、公寓

Location: Guangzhou, Guangdong Province
Client: Franshion Properties (China) Limited
Site Area: 45 200 m^2
Building Area: 100 200 m^2
Function: Office, Commerce, Apartment

城市里的院落
A COURTYARD IN THE CITY

设计将高层建筑与城市空间相融合，打造人与自然和谐共处、建筑与自然和谐共生的生活空间。

In the design of this project, the high-rise building is integrated with the urban space, creating a living space where people and nature live together in harmony, and where buildings and nature live together in harmony.

广州南沙金茂湾项目集酒店、总部基地、写字楼、企业 CEO 商墅、复式公寓、J·LIFE 风情商业街、滨海洋房七大业态于一体。

规划设计基于对传统院落空间围合的研究，采用“错动”的形式布置功能流线，通过商业串联起写字楼和公寓，形成公私混合的城市空间院落。

建筑由办公、商业和公寓三大功能区块组成， 公寓和写字楼采用“W”形的建筑布局，构成良好的自然采光条件。

以渐变式的像素为单元，对空间节点场地进行划分，使建筑、场地与景观形成有机整体，同时创造出丰富、多样的园区空间。

Nansha Jinmao Bay in Guangzhou is a complex of hotels, headquarters bases, office buildings, corporate CEO villas, duplex apartments, J-LIFE-style commercial streets , and seaside bungalows.

During the planning and design of the project, based on the study of the traditional courtyard space enclosure, the functional flow is well-arranged, linking the offices and apartments through the commercial chain, forming a mixed public and private urban space courtyard.

The building consists of three main functional blocks, namely office, commercial, and apartment. The apartments and offices are laid out in a "W" shape to provide optimum natural light.

The spatial node sites are divided into units of graduated pixels, so that the buildings, sites, and landscape form an organic whole while creating a rich and varied park space.

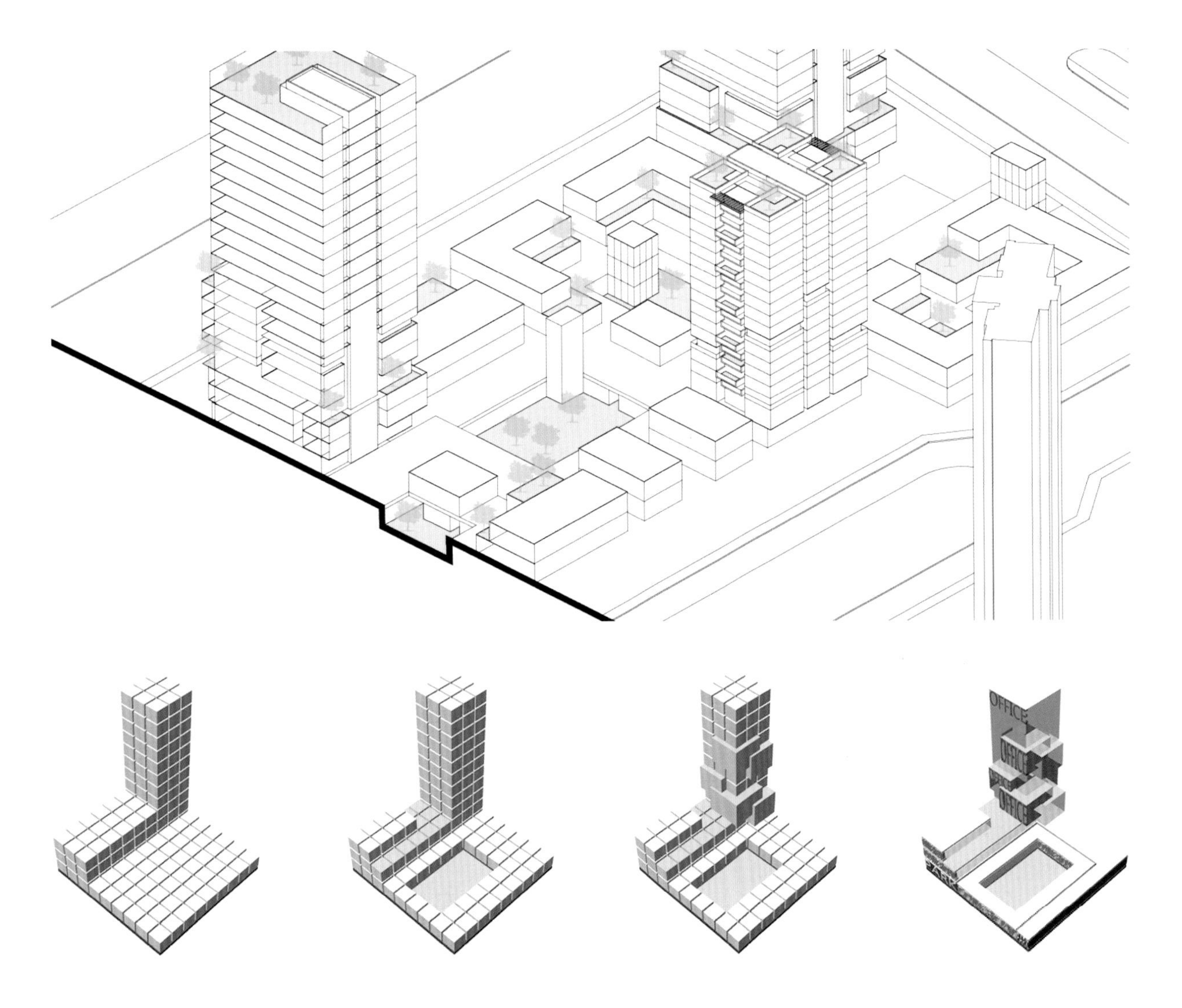

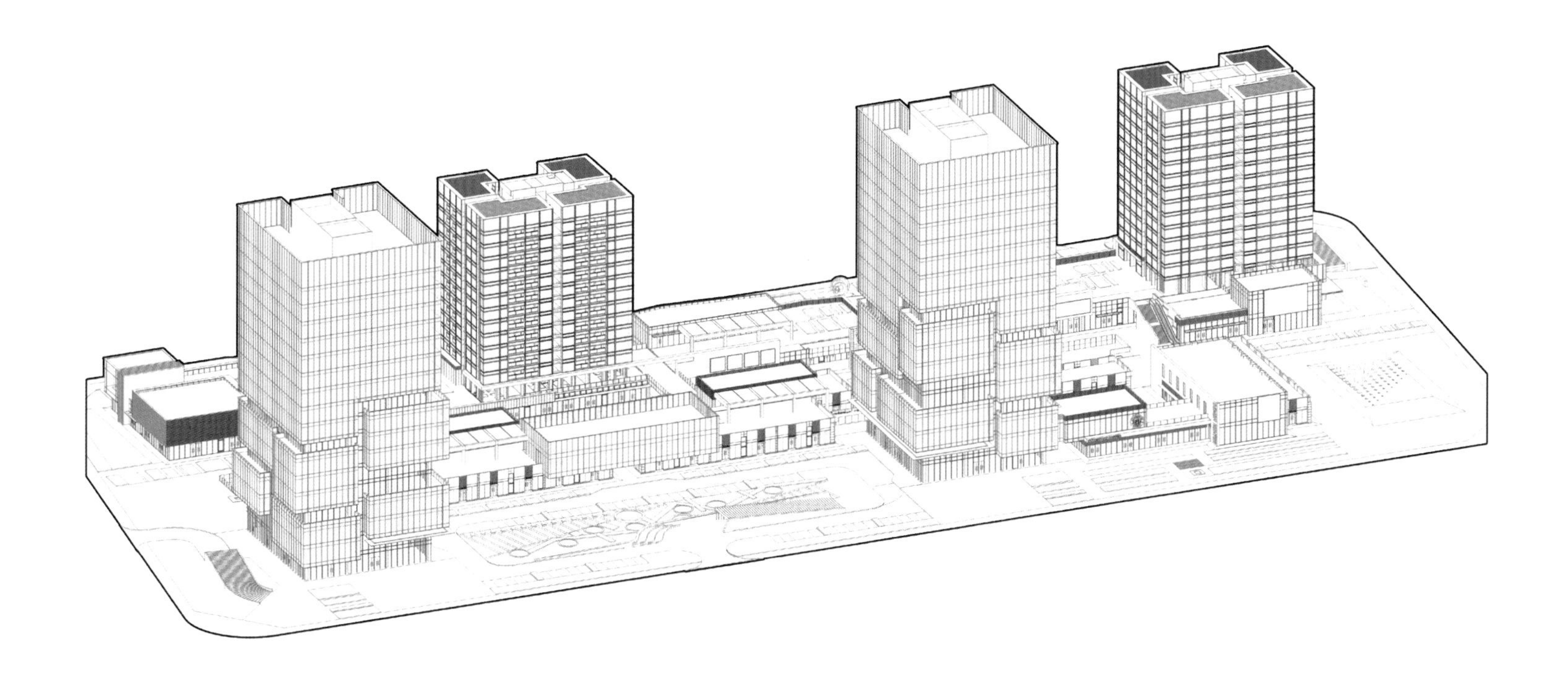

阜阳·北外附属新华外国语学校

XINHUA FOREIGN LANGUAGES SCHOOL AFFILIATED TO BFSU · FUYANG

位　　置：安徽省阜阳市
客　　户：新华地产投资有限公司
用地面积：16.48 万 m^2
建筑面积：12 万m^2
功　　能：学校建筑

Location: Fuyang, Anhui Province
Client: Xinhua Real Estate Investment Co., Ltd.
Site Area: 164 800 m^2
Building Area: 120 000 m^2
Function: School Building

花园式综合学校

GARDEN-STYLE COMPREHENSIVE SCHOOL

营造一片宁静、平和的净土，给孩子们充足的绿地、绿树和开阔的天空。

This project aims to create a peaceful, calm, and clean land where children can enjoy ample green space, green trees and open skies.

学校是求知的地方，“？”给了我们设计的启发，体育馆做成“？”形态，采用坡地地景消隐建筑，聚焦森林校园绿化中心。

规划以人行步道为轴线，以森林绿化为中心，把整体学校分为五区一中心布局形态。建筑从南往北慢慢升起，南部建筑屋顶与地面绿化融合一体，北部自然形成室外攀岩基地，从而成为森林校园中心。

在这里，对建筑起主导作用的因素是尺度，场地的关键特性是视线与人的行走路径。贯穿校区的步行道路中用起伏的坡地、台阶和折形坡道进行组合，创造出吸引孩子们嬉戏、读书、交往的空间。

各教学区平面布局采用“院落组团+内外廊+变化丰富屋顶”形式，以院落为组团，组织各教学功能，形成既独立又相互关联的形式空间。

A school is a place for seeking knowledge. Inspired by the symbol "?", we made the gymnasium in the shape of a half-arc question mark "?", while using a sloping landscape to conceal the building, thus achieving the effect of focusing on the green center of the forest campus.

In the planning process, we have divided the school into five zones and one center, using the pedestrian walkway as the axis and the forest greenery as the center. The buildings rise slowly from south to north, with the roofs of the buildings in the south integrating with the greenery on the ground and the outdoor rock climbing base in the north, thus forming the forest campus center.

Here, the dominant factor influencing the buildings is scale, and the key characteristics of the site are sight lines and human walking paths. The combination of undulating slopes, steps, and winding ramps in the footpaths that run through the campus create spaces that attract children to play, read, and socialize.

The layout of each teaching area is in the form of Courtyard Clusters + Internal and External Corridors + Varied Roofs", with the courtyards as clusters to organize the teaching functions and form a formal space that is both independent and interrelated.

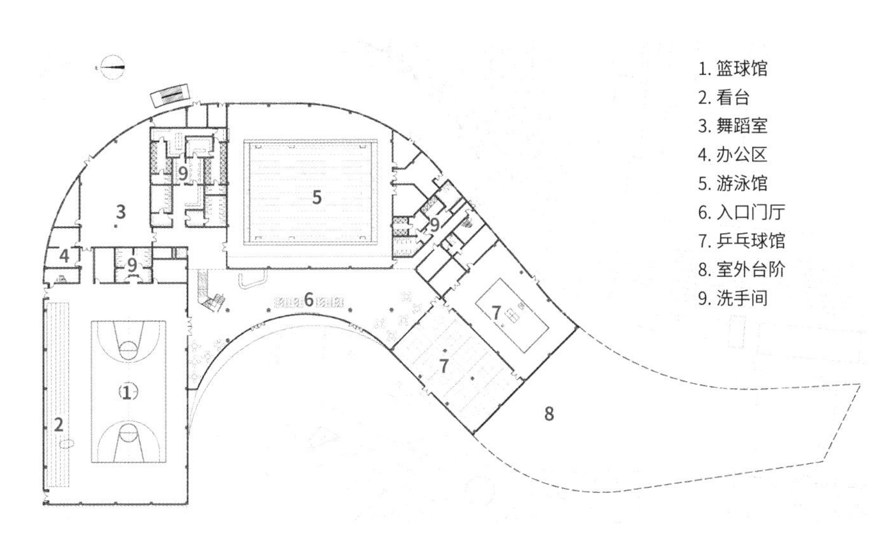
1. 篮球馆
2. 看台
3. 舞蹈室
4. 办公区
5. 游泳馆
6. 入口门厅
7. 乒乓球馆
8. 室外台阶
9. 洗手间

元生

贵阳·中铁阅花溪九年制中小学

YUEHUAXI NINE-YEAR MIDDLE AND PRIMARY SCHOOL OF CREG · GUIYANG

位　　置：贵州省贵阳市
客　　户：中铁置业集团贵州有限公司
用地面积：5.96 万 m^2
建筑面积：2.85 万 m^2
功　　能：学校建筑

Location: Guiyang, Guizhou Province
Client: China Railway Engineering Group (Guizhou) Co., Ltd.
Site Area: 59 600 m^2
Building Area: 28 500 m^2
Function: School Building

求知长廊 书山叠院

KNOWLEDGE CORRIDOR AND BOOKS COURTYARD

方案采用“新型聚落式校园”的设计理念，希望塑造更加自主、开放、多方位共享的教学空间，打造更具独特性、标志性、公共性的新型校园，实现学生在自然中学习的理念。

The scheme adopts the design concept of "new settlement campus", hoping to create a more autonomous, open, and multi-faceted shared teaching space, creating a new campus that is more unique, iconic, and communal, and realizing the concept of students learning in nature.

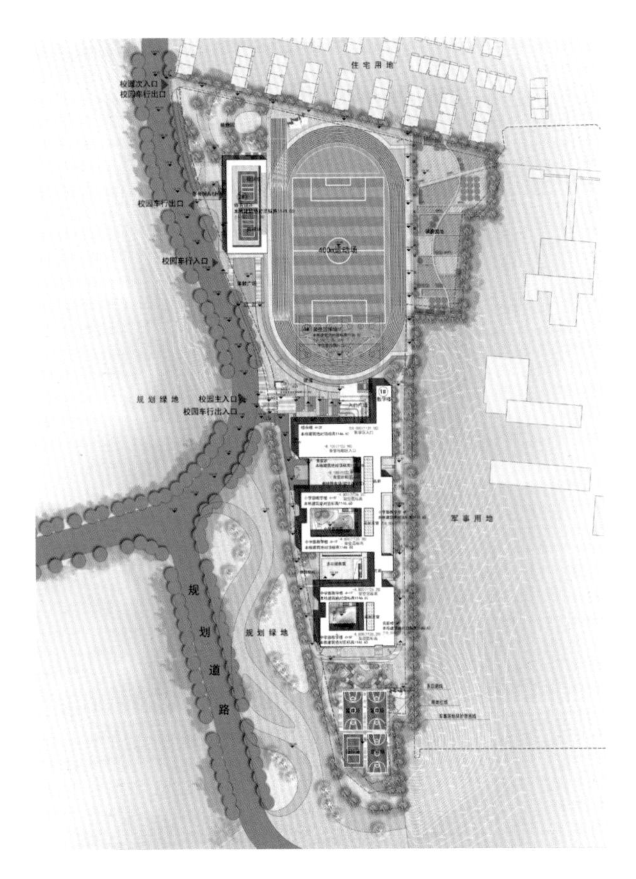

学校设计采用“一廊、一厅、两重院”的概念思路，打造出求知长廊作为贯穿校园的骨架结构，串联起学校的各个功能空间及趣味空间，方便师生使用，也为师生的交流创造了更多的机会。小学部与初中部相互独立，均采用围合书院的形式，书院底层架空处理，结合景观及绿化设计为学生创造了大量的交往及感知自然的机会。开放共享、串联互动、绿色生态的教学空间设计，为学生创造一个在自然中开心学习的“教育领地”。立面设计简洁而富有变化，通过标准化构件的排列组合及鲜明色彩的应用，打造出了一个独具标示性而又质感亲切的校园。

In the design, the school adopts the concept of "one corridor, one hall, and two courtyards", creating a knowledge corridor as the cage construction of the campus, which efficiently and conveniently connects the various functional and interesting spaces of the school, making it convenient for teachers and students to use and creating more opportunities for them to communicate. The primary and junior sections are managed independently of each other, and each adopts the form of an enclosed books courtyard, with the ground floor of the books courtyard being elevated and combined with the landscape and greenery design to create plenty of opportunities for students to interact and feel nature. The open, shared, interactive, green, and ecological design of the teaching space creates an "educational territory" for children to learn happily in nature. The facade design is simple and varied, with the arrangement and combination of standardized components and the use of vibrant colours to create a distinctive and intimate campus image.

明德厚学 求是创新

贵阳·观山湖第八中学
GUANSHANHU NO.8 MIDDLE SCHOOL · GUIYANG

位　　置：贵州省贵阳市
客　　户：中铁置业集团贵州有限公司
用地面积：2.8 万 m^2
建筑面积：1.13 万 m^2
功　　能：学校建筑

Location: Guiyang, Guizhou Province
Client: China Railway Real Estate Group (Guizhou) Co., Ltd.
Site Area: 28 000 m^2
Building Area: 11 300 m^2
Function: School Building

如同有机生命体般自由生长
GROWING FREELY LIKE AN ORGANIC LIFE FORM

建筑犹如一棵大树，负隅向前，这是一所关于自然和生长的学校。

The building is like a big tree, moving forward with indomitable will. It speaks of a school about nature and growth.

学校用地为东北—西南走向的狭长不规则地块，为了突破用地的限制，建筑采用了类似树枝的空间结构。

建筑犹如大树一般，躯干紧靠高差近 16 m 的峭壁，向着阳光与空气不断延展。建筑主体南向和东南向的建筑作为常规教室，西向和西南向的建筑作为合班教室及其他辅助功能空间。

风雨操场和食堂放置在地块的西南角，其屋顶被设置为运动场。运动场和教学楼的西南向“分枝”连成一体，使得屋面起起伏伏，又似节节爬升。教学空间由室内延伸到户外，自然景观由各层屋面渗透到室内，形成一个一体化、自由的室内外无界限的立体生态校园。

The school land is a long and narrow irregular plot with northeast-southwest trend. In order to overcome the limitation of land use, the building adopts a spatial structure similar to tree branches.

The building is like a big tree, with its trunk close to the cliff with a height difference of nearly 16 meters, extending towards the sun and air. The south and southeast "branches" of the main body of the building have become regular classrooms, while the west and southwest "branches" have become classrooms and other auxiliary functional spaces.

The rain proof playground and canteen are located in the southwest corner of the plot, and the roof is set as the playground, which is connected with the southwest branch of the teaching building, making the roof rise and fall. The teaching space extends from the indoor to the outdoor, and the natural landscape penetrates from the roofs of each floor to the indoor, forming an integrated and free three-dimensional ecological campus with no boundary between indoors and outdoors.

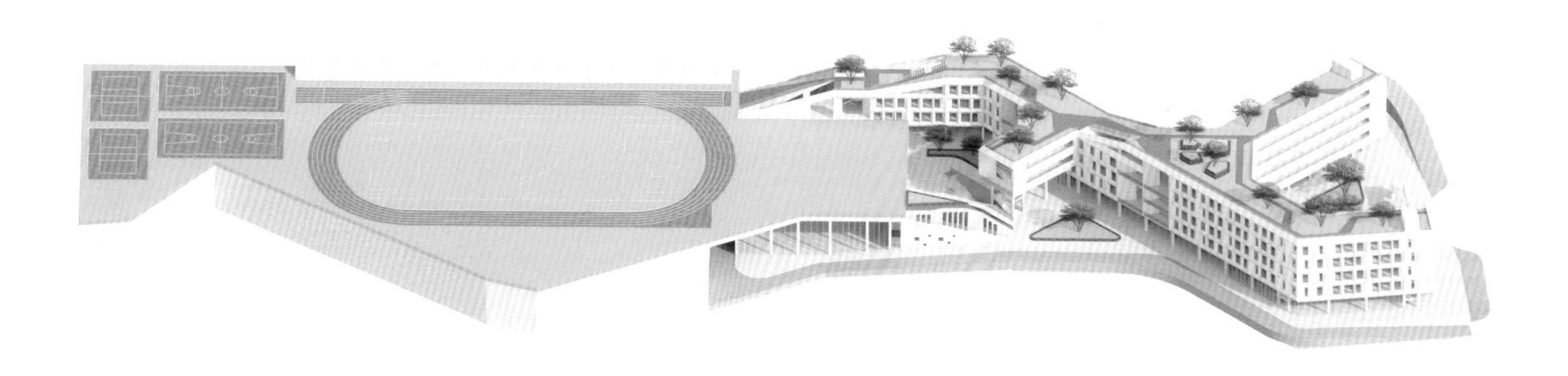

深圳·宏发悦见公园里
HONGFA YUEJIAN PARK · SHENZHEN

位　　置：广东省深圳市
客　　户：深圳市宏发投资集团有限公司
用地面积：3.08 万 m^2
建筑面积：22.88 万m^2
功　　能：超高层办公、超高层住宅、商业、人才房

Location: Shenzhen, Guangdong Province
Client: Shenzhen Hongfa Investment Group Co., Ltd.
Site Area: 30 800 m^2
Building Area: 228 800 m^2
Function: Super High-Rise Office, Super High-Rise Residence, Commerce, Talent Housing

云卷云舒，赴一场城市里的浪漫之约
CREATE A ROMANTIC URBAN SPACE WITH THE FORM OF "CLOUDS ROLLING AND STRETCHING OUT"

“云卷云舒”，四维变换的横向曲线创造出流动感的立面形象，形成城市天然的焦点。

With the form of "clouds rolled and stretched out", the transverse curved surface of four-dimensional transformation creates a flowing facade image and form the natural focus of the city.

作为 180 m 的超高层办公建筑，设计在整体横向线条的基础上于第一个避难层的位置创造了极富有编织感的线条变化。舒缓的流动曲线，加强了城市景观轴带的延展性，丰富了视觉层次。

裙房沿用塔楼银白色铝合金打造，一体化的线条设计，精致时尚。商业入口舒缓内退，通过视点延伸引导人群内聚，富有律动的曲线结合灯光打造个性化且富有吸引力的商业空间。

超高层住宅采用圆润精致的现代风格，灵动轻巧，别具一格。小曲线设计也与建筑整体统一。

整体项目形象统一和谐，横向云端舒展，集多种复合功能于一身，形成工作居住皆宜的城市新社区。

As a 180-meter high-rise office, the design of this project creates a very woven line change from the position of the first evacuation floor on the basis of the overall horizontal line. The soothing flowing curve enriches the visual level while strengthening the extension of the urban landscape axis.

The podium is made of the same silver-white aluminum alloy as the tower, and the integrated line design is exquisite and fashionable. The commercial entrance is gently receding, leading the crowd to gather through the extension of the viewpoint, and the rhythmic curve combined with lighting creates a personalized and attractive commercial space.

The super high-rise residence adopts a sleek, exquisite modern style, which is nimble, light, and distinctive, and the small curve design is also unified with the office as a whole.

Under the condition of high capacity and high density, the overall image of this project is unified and harmonious, stretching horizontally to the clouds, integrating a variety of composite functions, forming a new urban community suitable for work and living.

深圳·龙华合正观澜汇三期
LONGHUA HEZHENG GUANLANHUI (PHASE III) · SHENZHEN

位　　置：广东省深圳市
客　　户：深圳市合正房地产集团有限公司
用地面积：7.14 万 m^2
建筑面积：61.24 万 m^2
功　　能：办公、住宅、商业、产业研发

Location: Shenzhen, Guangdong Province
Client: Shenzhen Hazens Real Estate Group Co., Ltd.
Site Area: 71 400 m^2
Building Area: 612 400 m^2
Function: Office, Residence, Commerce, Industrial Research and Development

高容高密下的自然宜居城
A LIVABLE RESIDENCE WITH HIGH FLOOR AREA RATIO AND HIGH DENSITY

融合观澜山水印象，与自然对话，塑造场所记忆，打造集生活、产业、购物、生态、健康为一体的复合型超级社区。

This project aims to integrate the impression of Guanlan water, dialogue with nature, shape the memory of place, and create a complex super community integrating life, industry, shopping, ecology, and health.

设计通过优化城市界面与人居空间，解决“高容高密”的难题。

植入多层次绿地空间，弥补高容高密社区生态环境的缺失；引入多元配套功能，改善社区空间品质，构建老、中、青全龄化功能体系；搭建复合交通系统，提高社区交通使用效率。

在立面上，项目追求现代简约的设计手法，削弱了高容高密环境下带来的空间压迫感，降低了立面成本。在沿观澜大道的城市主界面上，设计通过一体化的设计手法，让塔楼裙房形成颇具动感时尚的城市意象，高山流水，一气呵成。

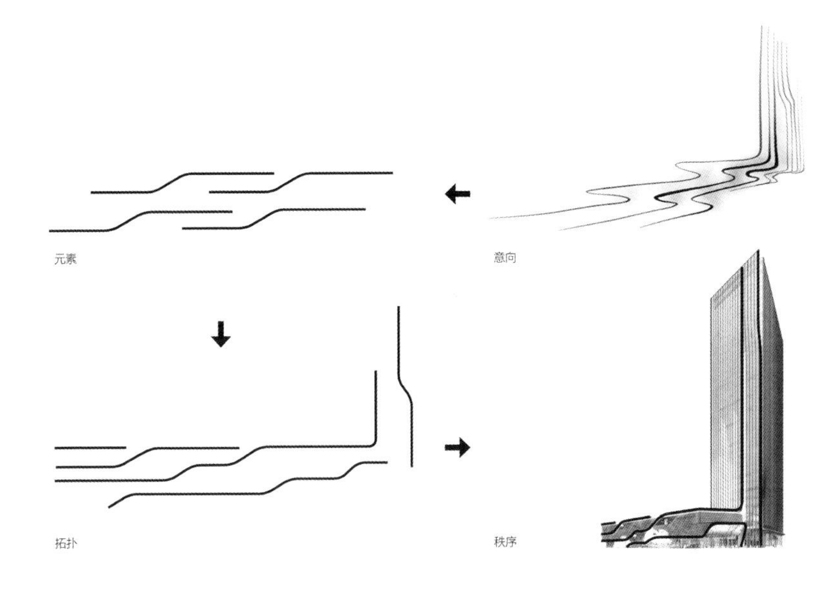

This project adopts a design strategy to optimize the urban interface and human living space in response to the challenge of "high floor area ratio and high density".

The features of this project are as follows: implanting multi-level green space to make up for the lack of ecological environment in high floor area ratio and high density communities; introducing multiple supporting functions to improve the quality of open space, building a functional system for people of all ages to form a comprehensive participation place for the elderly, middle-aged and youth; and building a complex transportation system to improve the efficiency of transportation.

Regarding the design of the facade, this project pursues modern and minimalist design techniques, which weakens the space pressure brought by high floor area ratio and high density, and also reduces the facade cost. Regarding the main interface of the city along Guanlan Avenue, the design of the tower podium is integrated to form a dynamic and fashionable urban imagery with a compact and coherent shape.

珠海 · 建发央璟

JIANFA YANGJING · ZHUHAI

位　　置：广东省珠海市
客　　户：建发房地产集团有限公司
用地面积：1.97 万 m^2
建筑面积：8.02 万 m^2
功　　能：住宅

Location: Zhuhai, Guangdong Province
Client: C&D Real Estate Co., Ltd.
Site Area: 19 700 m^2
Building Area: 80 200 m^2
Function: Residence

香洲正脉、中式华宅

AN AUTHENTIC CHINESE-STYLE LUXURY RESIDENCE IN XIANGZHOU

尊重城市空间肌理，“一轴一园，一带一府”，塑造荣耀名门府邸。

This project respects the spatial fabric of the city to shape the glorious mansion through the design of "one Axis, one Garden, one Belt, and one Mansion".

建發央璟

整体规划设计尊重城市空间肌理，建筑布局东西一字展开，呈现“一轴一园，一带一府”结构。“轴”即传统人文景观轴，借鉴传统建筑礼制空间序列，通过起承转合手法打造三进府邸院落空间；“园”为传统与现代结合，江南与岭南相融之园林；“带”是生态健康之带，东西一字展开；“府”是名邸雅府，尊享多维入户体验，精致贴心服务。

建筑形象沿城市东西线营造出高低错落的天际线形象，建筑外观结合传统皇家建筑要素及现代建筑技艺，塑造尊贵荣耀的名门府邸形象。建筑色调以霞红色配以古铜色，营造高雅建筑形象。置身府邸之中，匠心随处可触。

The overall planning and design of this project respect the spatial texture of the city, and the building layout is spread in an east-west direction in a single line, presenting the structure of "one Axis, one Garden, one Belt and, one Mansion". The "Axis" is the traditional humanistic landscape axis, which draws on the spatial sequence of royal architectural rituals to create a three-hall mansion courtyard space through the method of opening, developing, changing and, concluding; the "Garden" is a garden that combines tradition and modernity, and blends Jiangnan and Lingnan; the "Belt" is the ecological health belt, which spreads in an east-west direction in a single line; and the "Mansion" is a prestigious residence, which provides users with a dignified multi-dimensional entry experience and exquisite and intimate services.

The building is built along the east-west direction of the city to create a pleasing undulating skyline, and the exterior of the building combines the elements of traditional royal architecture and modern architectural techniques to create a prestigious and glorious image of a prestigious residence; the color palette of the building is red with dark curry and bronze to create an elegant architectural image and the ingenious details can be touched everywhere you are in the residence.

青雲書院

北京·远洋天著春秋
SINO-OCEAN TIANZHU CHUNQIU · BEIJING

位　　置：北京市
客　　户：远洋集团控股有限公司
用地面积：11.66 万 m^2
建筑面积：15.17 万 m^2
功　　能：住宅

Location: Beijing
Client: Sino-Ocean Group Holding Limited
Site Area: 116 600 m^2
Building Area: 151 700 m^2
Function: Residence

八大处畔，西山源脉
LIVE BY THE BADACHU PARK AND ENJOY THE SOURCE OF THE XISHAN MOUNTAIN

设计回归时光深处的大院生活，悉心拿捏“人、建筑、自然”三者之间的合适尺度，创新地传承了积淀千年的家族院落文化。

The design of this project returns to the courtyard life in the depths of time, carefully pinpointing the appropriate scale between "people, architecture and nature", and inheriting the family courtyard culture that has been accumulated for thousands of years in an innovative way.

项目拥有西山板块独一无二的禅林胜境，八大处、法海寺千古禅思萦绕，使之拥有“近于市，隐于林”的意境。规划采用外高内低的手法，使内部形成独立格局，体现“深宅大院，轴线分明”的理念，顿生“悠然见南山”之感。

This project is located in the unique Zen Forest in Xishan Mountain, close to Badachu Park and the Fahai Temple, where you can enjoy the ancient Zen thoughts and an artistic conception of "being close to the city and hidden in the forest". The planning adopts the technique of high outside and low inside to make the interior form an independent pattern, reflecting the concept of "a deep house and courtyard with clear axes" and creating a sense of "leisurely seeing Nanshan".

项目提取并运用花格窗、云纹图案、菱形垂鱼等丰富的中国古典建筑元素，局部搭配金属板装饰以及大面积的开窗，并运用强烈的虚实对比等现代建筑手法，将古典与现代的元素融为一体。单体建筑屋顶形式，屋架结构，斗拱制作，门窗装修，柱与檐口高宽比，均达到了力、曲线和美的统一。

This project abstractly extracts and uses rich Chinese classical architectural elements such as lattice windows, cloud patterns, diamond-shaped pendant fish, etc., and partially decorates with metal panels. Further, through the large-area window design and modern architectural techniques such as strong contrast between reality and imaginary, it perfectly blends traditional and modern elements. The roof form, roof truss structure, bucket arch making, door and window decoration, and height-width ratio of the column to the cornice of a single building all achieve the unity of force, curve and beauty.

宁德·春风里
CHUNFENGLI OF CATL · NINGDE

位　　置：福建省宁德市
客　　户：宁德时代新能源科技股份有限公司
用地面积：4.49 万㎡
建筑面积：8.52 万㎡
功　　能：住宅

Location: Ningde, Fujian Province
Client: Contemporary Amperex Technology Co., Limited
Site Area: 44 900 m^2
Building Area: 85 200 m^2
Function: Residence

当代“诗意栖居”语境下的定制化法式住宅
CUSTOMIZED FRENCH-STYLE RESIDENCE IN THE CONTEXT OF CONTEMPORARY "POETIC DWELLING"

项目通过“自由平面与古典立面的有机结合”“去繁就简下的经典范式萃取”以及“材料与构造细节的匠心推敲”，实现创新法式住宅的营造。

This project aims to create innovative French residences through the "organic combination of free plane and classical facade", "extraction of classical paradigm under the simplification of complexity" and "elaboration of materials and construction details".

春风里法式住宅在平面设计阶段，便以调查表的形式了解到不同业主对于住宅多样化的使用需求，并通过模型的推演找到法式古典语言的相关构图秩序，营造自由平面下的法式古典建筑立面。

在以法式建筑形态为参照的基础上，每一个部品设计都萃取法式 建筑元素之精华并融于一个有机整体中。我们采取的策略是去繁就简，克制任意堆砌线脚和细节，在每个部品的比例、样式、功能性乃至施工方式上均做精细化推敲，使得不同大小的住宅均可作为一个法式经典范式。

In the graphic design stage of the French-style residence Chunfengli, we learned about the needs of different owners for the use of the residence in the form of a variety of questionnaires, and found out the relevant composition order of the classical French elements through the deliberation of models, thus forming the French classical facade under the free plane.

Under the overall control of the French form, each component of this project is designed to extract the essence of classical French elements and integrate them into an organic whole. Our strategy is to simplify the complexity and refrain from arbitrary piling of lines and details, and to refine the proportions, style, functionality and construction methods of each component, so as to make each residence of different sizes a classic French-style model.

成都·交投电建国宾江山销售中心
CCIC – POWERCHINA REAL ESTATE GUOBIN JIANGSHAN SALES CENTER · CHENGDU

位　　置：四川省成都市
客　　户：中国电建地产集团有限公司
用地面积：0.5 万 m^2
建筑面积：0.089 万 m^2
功　　能：展示中心

Location: Chengdu, Sichuan Province
Client: PowerChina Real Estate Group Ltd.
Site Area: 5000 m^2
Building Area: 890m^2
Function: Exhibition Hall

一曲宋韵荡云间
A SONG RHYME SWINGS BETWEEN THE CLOUDS

借鉴宋代美学，以现代构成手法重新诠释美好的生活方式。

This project draws on the aesthetics of the Song Dynasty to reinterpret the beautiful lifestyle with modern compositional techniques.

项目结合场地特性，利用和开拓景观面和室内景观视野。蜿蜒小路，层层递进，打造可观、可行、可游的皈依之地。

设计通过对宋代建筑风格的重构，重新连接自然与文化空间的体验，整体强调平直线型感、轻柔纤巧感和构造精致感。建筑设计上，通过双重屋顶设计诠释了西南建筑的特点，对宋代建筑的要素进行设计优化，再现宋代美学。

引用古典造园手法，传承中式园林的空间秩序，将景观和空间贯穿，相互渗透，四重情境，和谐共生，做到传统与现代交融，演绎属于成都的宋式韵味。

By fully considering the characteristics of the site, this project is designed to maximize the use and development of landscape surfaces and indoor landscape views, and by designing winding paths in layers to create a place of conversion that can be viewed, walked and visited.

Through the reconstruction of modern Song language, the design of this project re-links the experience of natural and cultural space and emphasizes the sense of straightness, gentleness, and delicacy as a whole. In architectural design, it interprets the characteristics of southwestern Chinese architecture through the design of double roofs, preserving and optimizing the details of the architectural elements of the Song Dynasty, and finally reproducing the aesthetics of the Song Dynasty.

This project inherits the spatial order of Chinese gardens by introducing classical gardening techniques, and penetrates landscape and space through each other to create a harmonious coexistence of four scenarios, realizing the blend of tradition and modernity. Furthermore, this project combines modern design techniques to demonstrate the Song rhyme of Chengdu.

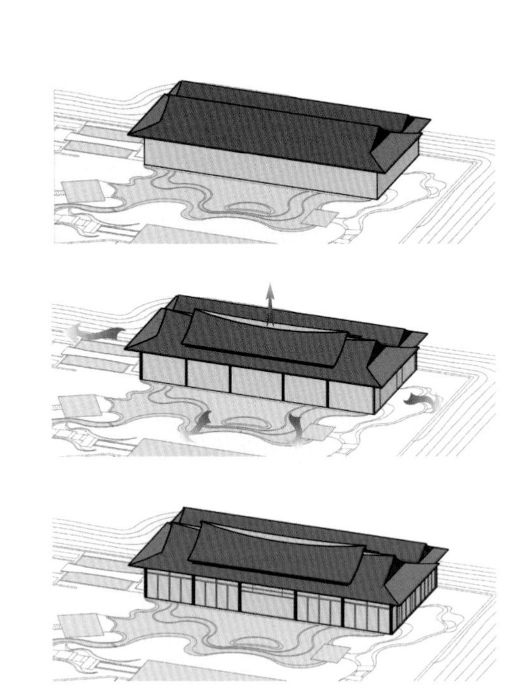

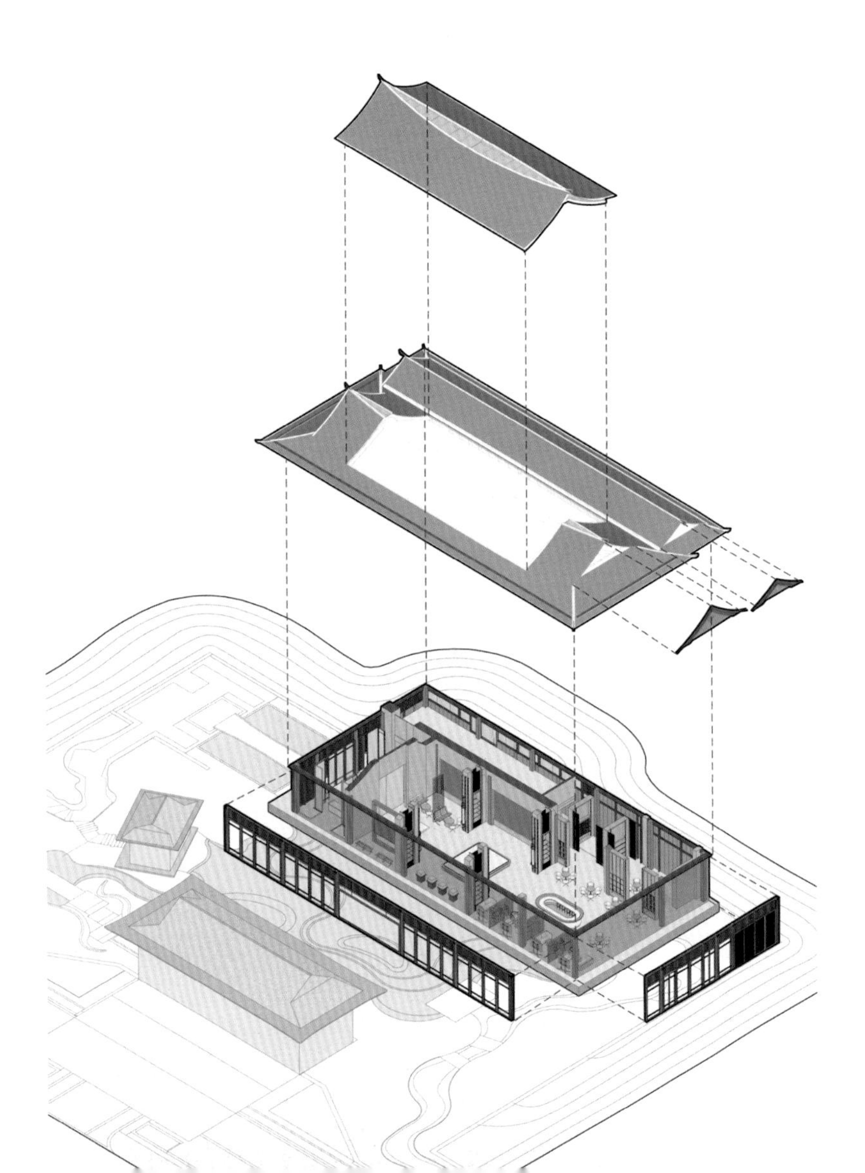

郑州·深业健康城
SHENYE HEALTH CITY · ZHENGZHOU

位　　置：河南省郑州市
客　　户：深业沙河（集团）有限公司
用地面积：36.6 万 m^2
建筑面积：122.34 万 m^2
功　　能：健康产业

Location: Zhengzhou, Henan Province
Client: Shenye Shahe (Group) Co., Ltd.
Site Area: 366 000 m^2
Building Area: 1 223 400 m^2
Function: Health Industry

“见山”，一种健康的生活方式，一张独特的健康名片
"SEEING THE MOUNTAINS", A HEALTHY LIFESTYLE AND A UNIQUE HEALTH BUSINESS CARD

设计抽取万山元素符号，为城市勾勒有机的空间聚落。通过创造与自然共生的空间场景，把展览馆塑造为当地一张独特的健康名片。

The design of this project extracts the symbols of Wanshan Mountain elements to outline the organic spatial settlement for the city. This project aims to shape the exhibition hall into a unique local health card by creating a spatial scene that is symbiotic with nature.

建筑形态结合南侧绿轴公园，引入渗透的绿谷。层峦叠翠式的山谷花园仿佛是从城市公园中自然生长而来。

跌宕起伏的流线组织，立体丰富的场景变化，为人们提供丰富的、自然健康的生活展示场景。

项目充分尊重当地的生态基底，秉持可持续、生态及人性化的设计理念，抽取“山、溪、谷、林”等代表自然健康的要素，以高度抽象的设计手法，化山为形，以溪为脉，退台为谷，植林作境。

The architectural form is combined with the Lvzhou Park on the south side, and the infiltrated greenery is introduced. The valley gardens with their layers of greenery seem to have grown naturally from the city park.

The ups and downs of streamlined design and three-dimensional rich scene changes provide people with rich, natural and healthy life scenes.

This project fully respects the local ecological base, adheres to the sustainable, ecological, and humanized design concept, extracts "mountains, streams, valleys, forests" and other elements that represent natural health, and uses highly abstract design techniques to build the building form according to the mountain situation, design the architectural vein along the stream, apply the set-back model design in the valley, and plant green forest to create a relaxed realm.

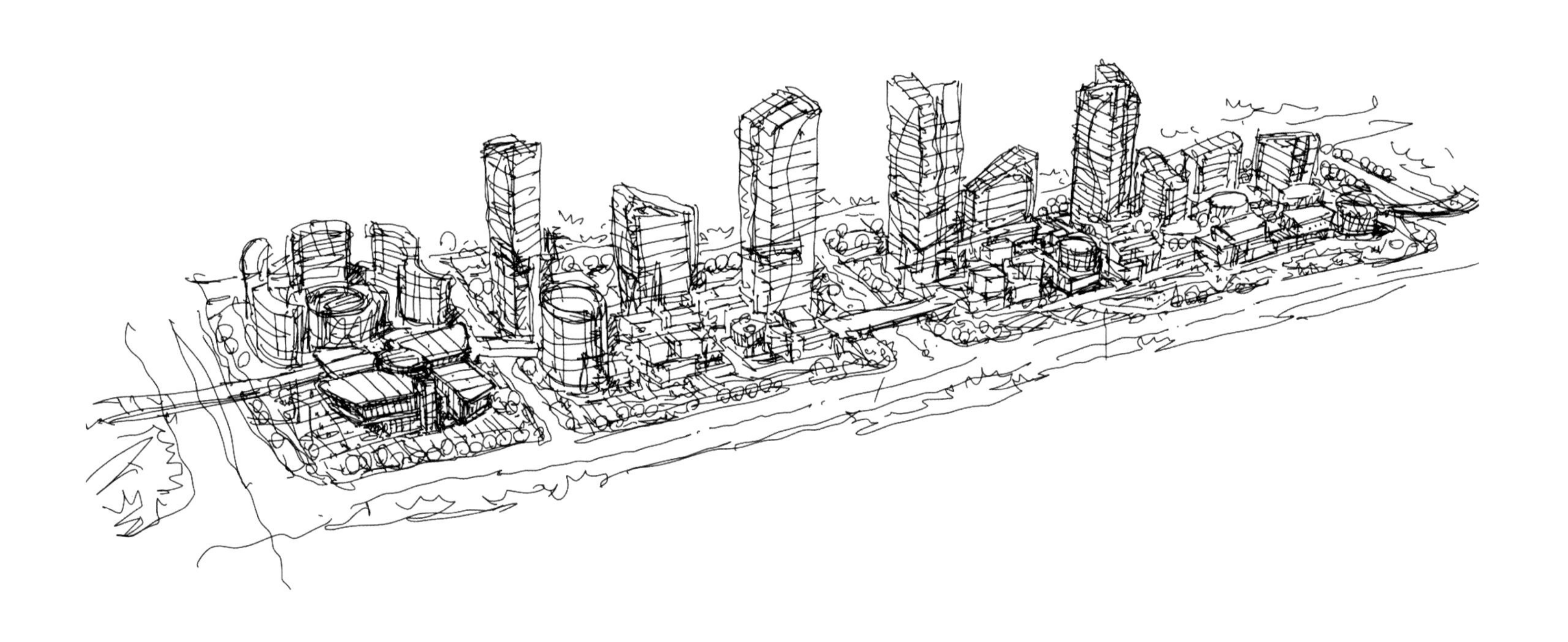

UNI
QLO

长沙·梅溪湖国际新城
MEIXI LAKE INTERNATIONAL NEW CITY · CHANGSHA

位　　置：湖南省长沙市
客　　户：方兴地产（中国金茂）
用地面积：59.02 万㎡
建筑面积：65.72 万㎡
功　　能：办公、商业、住宅

Location: Changsha, Hunan Province
Client: Franshion Properties (China Jinmao)
Site Area: 590 200 m^2
Building Area: 657 200 m^2
Function: Office, Commerce, Residence

低密度城市后花园
A LOW-DENSITY URBAN BACK GARDEN

以稀缺山水资源为依托，植入立体复合的生态体系，完成城市到自然的完美过渡，打造低密度的城市后花园。

Relying on scarce landscape resources, this project implants a three-dimensional compound ecological system, achieves the perfect transition from city to nature, and creates a low-density urban back garden.

方案希望设计两条贯穿龙王港河南北岸基地的空间景观主轴，串联梅溪湖、智慧公园、象鼻窝公园等重要城市空间，重点打造沿城市主干道展开的主要交通人流交汇节点，通过每个地块核心空间的视线景观次轴，将 7 块较为分散的城市地块从流线上、空间上、视线上均有机相连，和城市空间真正做到无缝对接。

项目包含生态高层办公区、水岸情景商务区、区域展示中心、滨水公共开放空间、湿地公园及低密度人居社区共六大片区。

The original design intention of this project is to design two spatial landscape spindles running through the base of the north and south banks of Longwanggang River, linking key urban spaces such as Meixi Lake, Zhihui Park, and Xiangbiwo Park, aiming to create major traffic and pedestrian intersection nodes along the main urban roads, and organically connect the seven scattered urban plots from the flow, space, and sight lines through the sight line landscape sub-axis of the core space of each plot to truly achieve a seamless connection with the urban space.

This project consists of six distinctive districts: ecological high-rise office district, waterfront business district, regional exhibition center, waterfront public open space, wetland park, and low-density residential community.

GUCCI
Armani
BOSCH

中山 · 华润马山片区规划项目
CR LAND MASHAN AREA PLANNING PROJECT · ZHONGSHAN

位　　置：广东省中山市
客　　户：中山市华润置地房地产发展有限公司
用地面积：55 万m²
建筑面积：159.57 万m²
功　　能：商业、住宅、文教

Location: Zhongshan, Guangdong Province
Client: China Resources Land (Zhongshan) Co., Ltd.
Site Area: 550 000 m²
Building Area: 1 595 700 m²
Function: Commerce, Residence, Culture and Education

岐江都心，万象名城
THE HEART OF THE QIJIANG NEW TOWN AND THE FAMOUS MIXC

项目依托当地自然景观与历史资源，打造出一个集旅游、商业、文化、生活于一体的 RBD 项目。

Relying on the local natural landscape and historical resources, this project creates an RBD project integrating tourism, commerce, culture, and living.

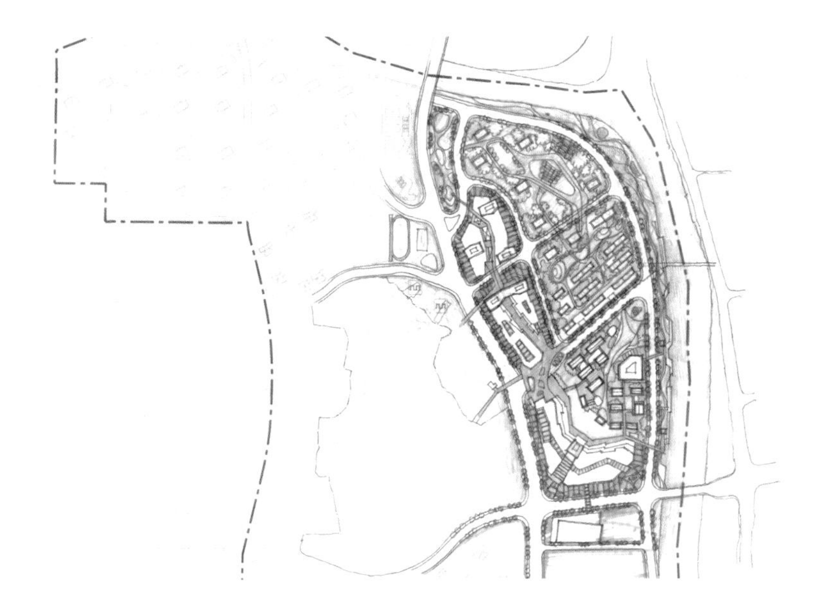

整体采用“一心一环一带”的规划结构，串连出“连城、游山、玩水”的功能业态。“山、城、水”三个魅力轴线共同构成了城市的空间廊道。

“一心”以万象城为核心，是整个区域的发展引擎，其中包括商业综合体、特色步行街等不同业态，为中山人民提供了丰富的购物体验；“一环”是由活动节点构成的山水乐游环，为区域提供了独特的旅游目的环线；“一带”联动了钟楼广场、会江广场两大城市公共空间，为中山市民打造出世界级的滨水公共场所。

In the overall planning of this project, a planning structure of "One Core, One Ring and One Belt" is adopted to link up the functions of "connecting the city, travelling in the mountains, and playing with the water", and the three attractive axes of "mountains, city and water" together create the spatial corridor of the city.

The "One Core", with MIXC as the core, is the development engine of the entire region, including commercial complexes, special pedestrian streets, and other different types of businesses, providing a rich shopping experience for the people of Zhongshan; the "One Ring" is a mountain and water pleasure ring composed of activity nodes, providing a unique tourism purpose ring road for the region; the "One Belt" links the two major public spaces of Zhonglou Square and Huijiang Square, creating a world-class waterfront public place for the people of Zhongshan.

成都・港中旅海泉湾度假区
HONG KONG CHINA TRAVEL SERVICE HAIQUANWAN RESORT · CHENGDU

位　　置：四川省成都市
客　　户：港中旅（中国）投资有限公司
用地面积：158.68 万㎡
建筑面积：77.41 万㎡
功　　能：住宅、文旅建筑

Location: Chengdu, Sichuan Province
Client: China Travel International Investment (China) Hong Kong Co., Ltd.
Site Area: 1 586 800 m^2
Building Area: 774 100 m^2
Function: Residence, Cultural and Tourism Building

悠游沱江乐水畔，宜居蓉城后花园
RELAXING ON THE RIVERSIDE OF THE TUOJIANG RIVER AND LIVING LIVEABLE IN THE BACK GARDEN OF CHENGDU

顺地就势，尊重地域文化，使建筑与环境相融合，寻找天人合一的惬意。

This project is in keeping with the terrain and respects the culture of the region, blending buildings with the environment to seek a pleasant harmony between man and nature.

项目位于四川省金堂县，规划结构上强调沿北河以及从泉眼到人工湖的两条轴线。在两条轴线上形成若干重要的空间节点，并使两条轴线互相联系，串联起整个规划中 6 种不同的功能组团。

建筑顺地就势，提取当地文化特色作为设计元素，既保持场地原生环境，又提高了项目地域文化底蕴，在适应成都市场的差异化特征的同时，满足了客群的体验型度假需求，打造出可住可游的生态型温泉度假小镇。

This project is located in Jintang County, Sichuan Province. In terms of planning structure, emphasis is placed on two axes along the North River and from the spring to the artificial lake. A number of important spatial nodes are formed along the two axes, and the two axes are linked to each other, weaving together six different functional groups in the entire plan.

The buildings are in keeping with the terrain and are designed in accordance with the local cultural characteristics as design elements, maintaining the original environment of the site, enhancing the cultural heritage of this project, adapting to the differentiated characteristics of the Chengdu market while meeting the experiential holiday needs of the clientele, and creating an ecological hot spring resort town that can be lived in and visited.

上海·诺港科学公园项目概念规划设计

CONCEPT PLANNING OF NUOGANG SCIENCE PARK · SHANGHAI

位　　置：上海市
客　　户：上海诺港科学集团有限公司
用地面积：32.96 万㎡
建筑面积：40.67 万㎡
功　　能：办公、产业研发、科学公园

Location: Shanghai
Client: Shanghai Nuogang Science Group Co., Ltd.
Site Area: 329 600 m^2
Building Area: 406 700 m^2
Function: Office, Industrial R&D, Science Park

无限循环的 LOOP
AN INFINITE LOOP

无限循环的 LOOP, 代表了对时间、空间、生命的思考与不断探索，是科学家的精神体现。

The infinite LOOP represents the thinking and continuous exploration of time, space and life, and is also the spiritual embodiment of scientists.

SIEMENS
DNKY

在整体规划上，将公园作为项目核心，融入“LOOP”精神，以自然为本，力求创造出一个融合生态、科技、艺术、人文的探索型科学社区。以循环的“LOOP”串联整个场地，结合多元的生态景观，与自然充分混织，形成高效互联的生态服务系统。

各种功能组团及公共空间有序的分布在整个系统中。综合研究楼的设计灵感来自于弦理论。中央科学公园以莫比乌斯环作为设计的元素，强调空间的连续和无限循环。科学研究组团的灵感来自于生物分子的特性。通过基础研究单元的无限复制、衍生，形成各具特色的研发组团。

Regarding the overall planning, this project takes the park as the core, incorporates the spirit of "LOOP", and takes nature as the foundation, striving to create an exploratory scientific community that integrates ecology, technology, art, and humanity. This project uses a circular "LOOP" design to link the entire site, combining diverse ecological landscapes and fully intertwining with nature to form an efficient and interconnected ecological service system.

In terms of layout, the functional clusters and public spaces are distributed in an orderly manner throughout the system. The design of the Integrated R&D Building is inspired by string theory. The Central Science Park uses the Möbius ring as a design element, emphasizing the continuity and infinite circulation of space. The scientific research clusters are inspired by the properties of biological molecules, while the distinctive R&D clusters are formed through the infinite reproduction and derivation of basic research units.

作品集项目分类列表

CATEGORY LIST OF PROJECTS IN THE COLLECTIONS

区域规划

REGIONAL PLANNING

居住社区

RESIDENTIAL COMMUNITY

文旅度假

TOURISM & VACATION

产业办公
INDUSTRIAL & OFFICE

文化教育
CULTURAL EDUCATION

体育公园
SPORTS PARK

会展展厅
EXHIBITION HALL

商业医疗
COMMERCIAL & HEALTHCARE

POSTSCRIPT: LET DREAM SHINE IN REALITY

n recent years, the ultimate philosophical questions such as "Who are you, where are you from, and where are you going?" are often asked everywhere. I am not sure whether you can answer the above questions in terms of PT Architecture Design (Shenzhen) Co., Ltd. (PTAD) after finishing reading this collection of works.

believe that it is not a tough thing to publish a couple of beautiful collections for any architectural design company of any size that has been operating in the Chinese design market for decades. But it must be not easy to fulfill most of these projects that the collections enclosed. Even though there are neither grand national projects nor speculative new projects in this collection, yet what inside our collection are mostly the "good neighbors" in the urban architectural group images or the "excellent tudents" who are always at the head of the line but never stand out. Whether this simpleness and truth is the annotation to the 25 years life of PTAD?

Twenty-five years ago, Peddle Thorp Architects originated from Australia and took root in Shenzhen, China. The outstanding architects and designers from Australia and other countries/regions, more from all over China, came to this frontier city of reform and opening up to join PT Design®, striving and growing together. Chinese society has experienced historic changes in the past few decades, the huge development energy has provided the architects with opportunities rarely seen in the world, while the architectural design industry has also experienced unique challenges in the world, which nearly redefined the ancient profession of architect in the construction tide in China.

The International Association of Architects published the Recommended Guidelines on Ethical Standards in 1998, which outlined the responsibilities and obligations of architects:

Firstly, the obligation of architects to the public: to abide by the law and to take full account of the social and environmental influence impacted by professional activities.

Secondly, the obligation of the architect to the clients: to practice faithfully and consciously, to consider technology and standards reasonably, to make judgements without any prevision or any prejudice, and prioritize academic and professional judgment over any other motivation.

Thirdly, the architect's duty to the profession: to maintain the dignity and quality of the profession, and to respect the legitimate rights and interests of others.

Fourthly, the obligation of architects to peers: to respect peers and to recognize their professional expectations, contributions and work results.

PT Architecture Design people may not recite silently these principles all the time, but we consciously try to adjust the above aspects between the occupation and the social environment with our simple cognition of architects in the design practice. We talk fewer about doctrines, but do more researches and pay more attention to quality, clients and social values. When your industry position is closer to the society, the diversity of social ideas, the difference between authority and public opinions, the constraints of technology and economy, and the grabbing of commercial interests, all will impact your professional standards and professional ethics more deeply. Fortunately, all these have cultivated PT Design® architects mature professional service consciousness rather than wear off our courage and specialty to create. We always believe that design can change lives and strive to make our dreams come true. Therefore, the works you can see from the collection today are filled with the joy of professional success and the satisfaction of contributing to the society, but permeated with unknown regret as well.

Compared with another collection of works compiled ten years after the establishment of PTAD, the projects presented in this collection of works for the 25th anniversary have clearer design concepts, richer design categories and more solid design quality. The PT Architecture Design keeps pace with the development of social economy, showing different characteristics in different stages and different fields. We no longer hold the aura of "luxury house design expert" in the field of residential architecture design, but pay more attention to the social attributes, market attributes and technical attributes of residential buildings, and create refined residential products with the concept of creating life by design. In the respect of protection and development of traditional buildings and blocks, we fully respect the local historical context. While protecting and utilizing old buildings, we boldly insert modern design elements to make many historic towns glow with the brilliance of modern life. In terms of the construction of stadiums and sports communities, we have built a bridge between famous sports operation institutions and the public with a series of wonderful innovative designs, achieving a win-win situation of social and economic benefits. In recent years, PT Architecture Design has been continuously engaging in large-scale public buildings. In addition to office buildings and hotels, we have also successfully completed and implemented two large exhibition projects with full professional design, including the Haikou Convention and Exhibition Center (EPC Phase II) and Changsha Convention and Exhibition Center, which have promoted our design and management quality greatly to a higher level.

Twenty-five years have passed in the blink of an eye. PT Architecture Design gradually becomes mature and colorful, and becomes a pivotal brand in the industry. Besides this collection of projects, there are the stories of countless imperfect, untimely, short existed, or even better or more innovative projects, and the stories of the passionate and hardworking designers behind them. Maybe all these should have been arranged in another collection.

While we are compiling this collection of works, the whole society is coming out of an unprecedented difficult moment. I would like to dedicate this collection to all the PT Design® people who are moving forward and sharing the difficult time.

Let dream shine in reality!

对比公司成立十年时的另一本作品集，本作品集所展现的项目，设计理念更加鲜明，设计品类更加丰富，设计质量更加扎实。柏涛建筑设计的步伐与社会经济的发展同步，在不同阶段、不同领域展示出不同的特色。在居住建筑设计领域，不再秉持“豪宅设计专家”的光环，而是更加专注住居的社会属性、市场属性和技术属性，以设计创造生活的理念，打造精细化的住宅产品。在传统建筑及街区保护与开发方面，充分尊重在地的历史文脉，在保护和利用古旧建筑的同时，大胆植入现代设计元素，使多个历史古镇焕发出现代生活的光彩。在体育场馆及运动社区营造方面，以精彩纷呈的系列创新设计，架起著名体育运营机构与社会群众共创健康体育社区的桥梁，取得了社会与经济效益的双赢。近年来，柏涛建筑设计向大型公共建筑出击，除办公楼与酒店建筑之外，还成功以全专业设计完成和实施工程管理了（EPC）海口会展中心二期和长沙会展中心两个大型会展项目，使公司的设计和管理质素向更高水准大幅迈进。

25 年，弹指一挥间。柏涛建筑设计逐渐走向成熟，走向丰盈，“柏涛建筑设计”已成为一个业界极具分量的品牌。这本作品集项目的背后，是无数个未曾完美，生不逢时，功亏一篑，或许专业上更加优秀或新锐的设计，以及项目背后那些充满激情和饱含辛劳的设计者的故事，也许应该另有一本作品集来承载。

此作品集编纂之际，全社会正在走出一个前所未有的艰难时刻，谨把这本作品集献给当下砥砺前行，共度时艰的全体柏涛人。

让梦想照进现实。

赵晓东

柏涛建筑设计董事、首席建筑师：赵晓东

Zhao Xiaodong

Director of PT Architecture Design, Chief Architect

PT DESIGN® CHINA

The Peddle Thorp Melbourne Pty Ltd is a world-renowned architectural practice, now known as China PT DESIGN since its expansion to China in 1998. In the past 3 decades, multiple design offices have been set up all across China in major locations such as Shenzhen, Shanghai, Beijing, forming an international design service network.

As a prestigious design practice and brand in China, PT DESIGN has established several subsidiary corporations including PTAD, PTBJ, PTL (Class A in the construction industry), PTI, PTLA, as well as Shanghai PT and its related companies. In each company, professional architects and engineers have provided design services for over 300 different clients, and have completed thousands of projects in multiple sectors, including mixed-use developments, residential, hotel, commercial, cultural, sports, leisure, office, and industrial buildings. PT DESIGN has won several awards awarded by professional authorities both within China and abroad.

PT DESIGN® provides full services in land acquisition, urban planning, architectural design, preliminary design, design development, construction documentation, as well as related services such as post-construction evaluation, marketing promotion, urban renewed properties, specialized engineering consultants, BIM design, and researches on prefabrication systems. With the motto "rooted in society, taking clients as a priority", PT DESIGN has a thorough understanding of the market system and constantly provides customers with innovative solutions. At the same time, PT DESIGN also offers special additional services, such as creating "realistic simulations and evaluations prior to construction", and "integration of the creative design with the Internet", both of which have been highly appreciated by clients.

In the past 3 decades, PT DESIGN has grabbed the opportunities of urban and rural developments all across China and is now stepping into a new era together with staff and clients. In line with national future strategies, PT DESIGN will continue to strengthen its teams and cooperate with the top designers around the world, building a world-class platform to amass creative talents. PT DESIGN will work harder to become a comprehensive service provider, in the leading of planning and architectural innovation, expanding to broader sectors of development and creating all industrial chains through upgrading the ability for creative services.

PT DESIGN ® , for a better living environment to create.

柏涛®设计中国机构

澳大利亚柏涛墨尔本建筑设计有限公司为全球知名建筑设计企业，1998年正式进驻中国以来逐步拓展为柏涛®设计中国机构，陆续设立深圳、上海、北京等一系列设计公司及香港办公室，逐渐形成了覆盖全国、连接世界的设计服务体系。

柏涛®设计中国机构以享有盛誉设计的服务于中国运营数十年，旗下包括柏涛建筑、柏涛北京、柏涛蓝森（建筑行业甲级）、柏涛国际工程顾问、柏涛景观，同时还包括上海柏涛及其相关的系列公司。各公司云集了大批知名建筑师和工程师，人才济济，至今服务客户超过300余家，完成的住宅、办公、酒店、商业、文化、体育、旅游、产业等项目多达数千个，并多次获得国际及中国权威机构的荣誉奖项。

柏涛®设计中国机构提供从土地获取顾问、城市规划、建筑方案设计、初步设计、施工图设计到设计总承包全程服务，以及项目后评估、市场推广、存量物业更新、专项工程顾问、BIM设计、建筑工业化体系研究等广泛的拓展服务。秉承“根植社会，客户为先”的理念，我们深刻了解市场需求，并不断为客户提供创新解决方案。同时，“先于建造的实景体验和评价”以及“互联网+设计创意一体化”是我们独具特色且广受客户好评的设计延申服务。

过去二十多年，我们幸运地把握了中国城乡开发巨大的机遇。而今我们带领员工、携手客户共同步入前所未有的新时代。配合国家未来战略，我们将不断强化自身团队，与国际建筑设计精英广泛合作，搭建具有国际一流水准的创意人才集群平台。升级创新服务能力，努力成为以规划及建筑设计创新为龙头的，贯通开发和营造全产业链的综合服务提供者。

柏涛®设计，为更美好的人居环境而创造。

PT ARCHITECTURE DESIGN (SHENZHEN) CO., LTD.

PT Architecture Design (Shenzhen) Co., Ltd. (PTAD) is one of the founding members of PT DESIGN® China and a cooperative agency of Peddle Thorp Melbourne Pty Ltd. It is a creative and service-oriented company based on China while targeting on the global market.

PTAD, together with PTL (Class A in the construction industry), PTI and PTLA, form the PT DESIGN® China company cluster in Shenzhen, providing comprehensive design services in all industrial chains for the construction market in China.

PTAD has experienced rapid developments since its establishment in 1998 and its business goes through all provinces and autonomous region/city/town, having successfully completed many outstanding and excellent projects, including planning, urban design, city complex, residential architecture, office buildings, tourist hotel resort, medical facilities, etc. In the fields of residential architecture upgrading, sports garden design and development, creation and updating of culture and touring town, etc., PTAD has leading design concepts and rich design experience with our dedicated team continuously and deeply taking special research. Design works of PTAD have been awarded multiple times by construction authorities at home and abroad.

PTAD absorbs talented people widely, and is dedicating itself to creating an international innovative platform for talents, gathering and working with many high-quality top designers and architects from China and abroad. There are many famous masters and continuous new staff. We always adhere to the principle of combining the advanced international design concept with the compliance with local regulations, as well as enablement of engineering construction and sustainability of the project usage in design, constantly presenting excellent design works to the China construction industry and the real estate market.

PTAD has been holding the design aim of taking priority in innovation, and being oriented in service. It has been building its reputation with unique design patterns, rich technical experience and sharp market observation. PTAD has established a market reputation and brand image deeply rooted in the public through many years' development, making itself a leading company in the field of planning and architectural design in China.

柏涛建筑设计（深圳）有限公司

柏涛建筑设计（深圳）有限公司（简称：柏涛建筑）是柏涛®设计中国机构（PT DESIGN）创始成员，是全球知名设计企业澳大利亚柏涛墨尔本建筑设计有限公司的中国合作机构。柏涛建筑立足中国、面向国际，是一家独具创意、注重服务的建筑设计公司。

柏涛建筑与深圳市柏涛蓝森国际建筑设计有限公司（建筑行业甲级）、柏涛国际工程设计顾问有限公司、深圳市柏涛环境艺术设计有限公司等共同组成柏涛®设计中国机构的深圳公司集群，为中国建筑市场提供综合性的全产业链设计服务。

自 1998 年成立以来，柏涛建筑发展迅速，在中国的业务范围遍及各省及自治区城乡，成功设计完成了许多令人瞩目的优秀项目，类型涉及规划、城市设计、城市综合体、居住建筑、办公建筑、旅游度假酒店、医疗设施等。在居住建筑升级与迭代、体育园区设计与开发、文旅小镇创建与更新等领域，有领先业界的设计理念和丰富的设计经验，并有专注的团队持续深入地进行专项研究。柏涛建筑的设计作品历年多次获得国际及中国权威机构的荣誉奖项。

柏涛建筑广纳人才，一直致力于打造国际化的创意型人才平台，聚集和联合了众多高素质的中外建筑师和设计精英，这里名师荟萃、新秀叠出。在设计中始终秉承将国际化的先进设计理念与本地规范的合规性、工程建造的可实施性及项目使用的可持续性相结合的原则，为中国建筑行业和房地产市场不断呈现优秀的设计作品。

“创新为先、服务为本”是柏涛建筑的设计宗旨。独特的设计思维、丰富的技术经验和敏锐的市场洞察是柏涛建筑的信誉之所在。历经多年的努力，柏涛建筑设计竖立起了广获赞誉的客户口碑和品牌形象，成为中国规划及建筑方案设计领域的领军企业。

柏涛® 历程
HISTORY OF PT DESIGN®

1998 柏涛建筑设计进入中国，落地深圳。

1999 深圳柏涛环境艺术设计有限公司成立。

2001 深圳卓越蔚蓝海岸竣工，以“海洋文化”为规划理念，成为住建部推荐的十大参观项目之一。

2003 深圳波托菲诺纯水岸竣工，华侨城集团借此项目首次发布“旅游地产”口号。

2003 深圳水榭花都竣工，首创“Townhouse 混合社区”、“两层超大空中花园”等建筑产品设计，对其后深圳豪宅产品带来深远而持久的影响。

2003 在上海设立分公司，上海柏涛建筑设计咨询有限公司成立。

2005 深圳泰格公寓获得 LEED-NC 认证，是国内首批获得 LEED 认证的商业项目。同年，泰格公寓列入国家建设部科技综合示范项目。

2006 深圳万科第五园竣工，在欧陆风盛行之时，以现代中式建筑开创了建筑界“中国风”。

2007 深圳香蜜湖 1 号竣工，作为城市中心的最后一块低密度用地和居住品质的象征，被人评价为“中国主流豪宅领袖”。

2008 深圳中信红树湾竣工，首创高层住宅空中花园建筑设计。

2010 北京华润橡树湾竣工。华润橡树湾产品系列的开篇之作，以学院派气质，开启了人文品质住区的先河。

2010 在北京设立分公司，柏涛建筑设计(北京)有限公司成立，完成了全国一线城市布局。

2011 由前奥运冠军体操王子李宁先生创建的全国首个公益性体育园——南宁李宁体育园开始试运营。此后，柏涛设计与李宁一路携手，开启全国系列李宁体育园的设计。

2013 黎阳 in 巷完工开放，构建黎阳古镇文化展示平台，助力黎阳老街成为 AAAA 景区。

2015 凝聚柏涛设计集体智慧的心路之作——《寻找心灵的尺度》一书正式发布。

2015 作为高品质滨海度假型第二居所的惠州华润小径湾竣工。

2016 具有建筑行业（建筑工程）甲级资质的深圳市柏涛蓝森国际建筑设计有限公司于深圳成立，柏涛设计开启全流程设计服务。

2018 以传承海运文化为出发点的滨海豪宅——深圳招商双玺时光道落成，备受瞩目。

2020 柏涛设计在众多知名设计机构参与的国际竞赛中脱颖而出，中标深圳深国际颐城栖湾里项目。

2021 柏涛设计担任设计总承包的海南国际会展中心二期竣工，为海南省重点项目，也是首届中国国际消费品博览会举办地。

2022 柏涛设计第一个国际超甲写字楼项目——武汉中海中心落成。

2023 约 200 m 高的深圳城市更新超高层地标性建筑——深圳帝豪金融大厦落成。

1998 PTAD entered China and ran business in Shenzhen.

1999 PTLA was established.

2001 COTEDAZUR in Shenzhen by Excellence Group was completed, featuring the planning concept of "marine culture", and was selected as one of the top 10 projects recommended by the MHURC to visit.

2003 Portofino Lake Garden in Shenzhen was completed, OCT Group announced the slogan of "Tourism Real Estate" for the first time.

2003 Shuixie Huadu Garden in Shenzhen was completed, which initiated the design of "Townhouse Mixed Community" and "Two-story Super Hanging Garden", exerting a far-reaching and lasting impact on the luxury residential products in Shenzhen.

2003 The Shanghai branch - Shanghai PT Architects was established.

2005 Shenzhen Taige Apartments was awarded LEED-NC certification, being the first commercial project in China to receive LEED certification. In the same year, Taige Apartments was listed as the comprehensive demonstration project of science and technology of the MHURC.

2006 The Fifth Park of Vanke in Shenzhen was completed, creating a "Chinese style" in the architectural world with modern Chinese-style architecture at a time when the European style was prevalent.

2007 Xiangmihu No.1 in Shenzhen was completed. As the last low-density site in the center of Shenzhen and the symbol of living quality, it was appraised as "the leader of mainstream luxury houses in China".

2008 CITIC Mangrove Bay in Shenzhen was completed, creating the first architectural design of high-rise residential hanging garden.

2010 CR Land Oak Bay in Beijing was completed. As the first product of CR Land Oak Bay Series, it creates the precedent of humanistic quality residential area with its collegiate temperament.

2010 The Beijing branch - PTAD · BJ was established. Until then, PTD completed its layout in the first-tier cities in China.

2011 Nanning Li-Ning Sports Park, the first public welfare sports park in China, founded by former Olympic champion Mr. Li-Ning, who was famed as the gymnastics prince, started trial operation. Since then, PT Design and Li-Ning have joined hands to start the design of Li-Ning Sports Park series products nationwide.

2013 The Liyang in Lane was completed and opened to the public, creating a platform to showcase the culture of Liyang Ancient Town and helping Liyang Old Street to become an AAAA scenic spot.

2015 The book *Searching for the Measure of the Heart*, a mental works that condenses PTD's collective wisdom, was officially released.

2015 The CR Land Xiaojingwan in Huizhou - a second high-quality coastal resort-was completed.

2016 PTL with Class A qualification of construction industry (architectural engineering) was established in Shenzhen, indicating that PTD started the full-process design services.

2018 Imperial Park of CMPD in Shenzhen, a coastal mansion based on the inheritance of maritime culture, was completed and attracted much attention.

2020 PTD won the bid for the Shenzhen International Yicheng Qiwanli Project in Shenzhen in an international competition with the participation of many well-known design agencies.

2021 The Hainan International Convention and Exhibition Center (Phase II), which PTD served as the design EPC, was completed. As a key project in Hainan Province, it was also the venue of the 1st China International Consumer Products Expo.

2022 Wuhan China Overseas Center, the first international super A office project of PTD, was completed.

2023 Shenzhen Regency Financial Building was completed, which is about 200 meters high and is a landmark building of super high-rise for city renewal in Shenzhen.

出 品 人：柏涛建筑设计（深圳）有限公司
名誉主编：王漓峰　赵晓东　吕学军
发起策划：柏涛建筑设计品牌中心
总 编 辑：滕怡
执行总编：欧阳霞
美术设计：姜静　郑佳英
责任校对：欧阳霞　姜静　陈桂明
英文翻译：北京好译来翻译中心

地址：广东省深圳市南山区华侨城中旅广场 华·生活馆 201A
电话：0755-26928866 / 26919976
传真：0755-2660 5399
邮箱：hr@ptadesign.cn
邮编：518000
网址：http://www.ptma.com.cn

图书在版编目（CIP）数据

柏涛建筑设计作品 : 2010-2023 / 柏涛建筑设计（深圳）有限公司编 . - 武汉 : 华中科技大学出版社 , 2023.5
ISBN 978-7-5680-9352-1
Ⅰ . ①柏… Ⅱ . ①柏… Ⅲ . ①建筑设计－作品集－中国－现代 Ⅳ . ① TU206

中国国家版本馆 CIP 数据核字 (2023) 第 072840 号

柏涛建筑设计作品 : 2010-2023
BOTAO JIANZHU SHEJI ZUOPIN:2010-2023
柏涛建筑设计（深圳）有限公司 编

出版发行：华中科技大学出版社（中国·武汉）
武汉市东湖新技术开发区华工科技园
电 话：（027）81321913
邮 编：430223
出 版 人：阮海洪

策划编辑：段园园
责任编辑：段园园 陈晓彤
版式设计：姜静 郑佳英
责任监印：朱玢

印　刷：深圳市国际彩印有限公司
开　本：889 mm×1194 mm 1/12
印　张：19.5
字　数：168 千字
版　次：2023 年 5 月 第 1 版 第 1 次印刷
定　价：268.00 元

華中出版

投稿方式：13710226636（微信同号）
本书若有印装质量问题，请向出版社营销中心调换
全国免费服务热线：400-6679-118 竭诚为您服务